THE OSPAR ARBITRATION
(IRELAND – UNITED KINGDOM)
AWARD OF 2003

PERMANENT COURT OF ARBITRATION AWARD SERIES

THE OSPAR ARBITRATION (IRELAND – UNITED KINGDOM) AWARD OF 2003

With an Introduction by Daniel Bodansky

Series Editor

BELINDA MACMAHON
Legal Counsel/Chief Editor
Permanent Court of Arbitration

T·M·C· ASSER PRESS

Published by T·M·C·Asser press
P.O. Box 16163, 2500 BD The Hague, The Netherlands
<www.asserpress.nl>

T·M·C·Asser press' English language books are distributed exclusively by:

Cambridge University Press, The Edinburgh Building, Shaftesbury Road,
Cambridge CB2 2RU, UK,
or
for customers in the USA, Canada and Mexico:
Cambridge University Press, 100 Brook Hill Drive, West Nyack, NY 10994-2133, USA
<www.cambridge.org>

ISBN 978-90-6704-295-6

The Permanent Court of Arbitration (PCA) is an intergovernmental organization, founded at the first Hague Peace Conference in 1899, and based in the Peace Palace in The Hague. It provides a wide range of dispute resolution mechanisms – including arbitration, conciliation and fact finding – in disputes involving various combinations of states, private parties and intergovernmental organizations. The PCA is committed to meeting the rapidly evolving dispute resolution needs of its member states and the international community.

International Bureau
Permanent Court of Arbitration
Peace Palace
Carnegieplein 2
2517 KJ The Hague
The Netherlands
Telephone: +31 70 302 4165
Facsimile: +31 70 302 4167
Email: bureau@pca-cpa.org
Website: www.pca-cpa.org

CONTENTS–SUMMARY

FOREWORD

This volume contains the Award of the Arbitral Tribunal established under the dispute resolution provisions of the 1992 Convention for the Protection of the Marine Environment of the North-East Atlantic ("OSPAR Convention"), to consider an international environmental dispute between Ireland and the United Kingdom which dates back to the early 1990s. The dispute concerns alleged potential radioactive pollution of the Irish Sea from a mixed oxide ("MOX") manufacturing plant at Sellafield, England. The *OSPAR* Arbitration was the first round of the notable *MOX Plant* Case.

Ireland commenced arbitral proceedings against the United Kingdom under two different treaties in 2001. Both arbitrations were administered by the Permanent Court of Arbitration ("PCA"). The first – the *OSPAR* Arbitration – was initiated in June of that year on the ground that the United Kingdom had failed to provide access to certain information regarding its decision to commission the MOX plant, in breach of Article 9 of the OSPAR Convention. The second arbitration – the *MOX Plant* Case – was brought under Annex VII of the United Nations Convention on the Law of the Sea ("UNCLOS") in October, and concerned the authorization and operation of the MOX Plant. The dispute also gave rise to proceedings in which Ireland requested provisional measures from the International Tribunal for the Law of the Sea pending constitution of the UNCLOS tribunal, as well as infringement proceedings brought by the European Commission against Ireland before the European Court of Justice.

The *OSPAR* Arbitral Tribunal was composed of Professor W. Michael Reisman (Chairman), Lord Michael Mustill PC, and Dr. Gavan Griffith QC. The PCA served as Registry for the arbitration. By agreement of the Parties, the hearings, held at the Peace Palace in October 2002, were open to the public, and the written pleadings, transcript of the proceedings and the Tribunal's procedural orders are available on the PCA website at <www.pca-cpa.org>.

The Tribunal was requested to address a comparatively narrow legal issue concerning information exchange between States. In its Final Award, rendered on July 2, 2003, the Tribunal, *inter alia*, rejected the United Kingdom's arguments regarding its jurisdiction over the dispute and the admissibility of Ireland's claims, but by majority decision denied Ireland's requests for information in its interpretation of the relevant provision of the OSPAR Convention. Appended to the Final

Award are a Declaration by Professor Reisman and a Dissenting Opinion by Dr. Griffith.

The PCA would like to express its gratitude to Daniel Bodansky, Woodruff Professor of International Law at the University of Georgia, for writing an astute and perceptive introduction on the contribution of the *OSPAR* Award to international law.

The Hague, September 2008

Christiaan Kröner
Secretary-General
Permanent Court of Arbitration

THE "OSPAR" ARBITRATION (IRELAND–UNITED KINGDOM)

Its Contribution to International Law

by

Daniel Bodansky*

I. INTRODUCTION

Not so long ago, concerns were commonly expressed about the infrequency of international dispute settlement. More countries were withdrawing from the compulsory jurisdiction of the International Court of Justice ("ICJ") than accepting it. Proposals for an international criminal court had languished for decades. Except for a few regional tribunals, such as the European Commission and Court of Human Rights – and a few specialized areas, such as investment arbitrations – international dispute settlement was a rarity.[1] In international environmental law, there was the *Behring Fur Seals* Arbitration[2] and the *Trail Smelter* Arbitration,[3] and not much else[4] – on average, barely a case a decade through the 1980s.

---

* Woodruff Professor of International Law, University of Georgia.

[1] See, e.g., J.G. MERRILLS, INTERNATIONAL DISPUTE SETTLEMENT (Cambridge University Press, 3d ed. 1998), at p. 165 ("by comparison with domestic courts, international courts and tribunals occupy a relatively insignificant position").

[2] *Behring Fur Seals* Arbitration (Gr. Brit./U.S.), Award, Aug. 15, 1893.

[3] *Trail Smelter* Arbitration (U.S. v. Can.), Award, Apr. 16, 1938; Award, Mar. 11, 1941.

[4] As Peter Sand noted in 1991: "[o]ne of the many myths of environmental law is the assumption – found in many legal textbooks – that international disputes in this field are settled along the lines of the 1941 *Trail Smelter* arbitration. Yet over the past 50 years, there have been only two inter-governmental dispute adjudications that could even remotely be compared to *Trail Smelter* – and even these claims (the 1957 *Lake Lanoux* arbitration between France and Spain and the 1968 *Gut Dam* arbitration between Canada and the USA) concerned classical questions of water use and flood damage, rather than a genuine environmental problem." Peter Sand, *New Approaches to Transnational Environment Disputes*, 3 INT'L ENVTL. AFF. pp. 193–206, at p. 193 (1991); see also Richard B. Bilder, *The Settlement of Disputes in the Field of the International Law of the Environment*, 144 RECUEIL DES COURS pp. 139–239 (1975) (noting that "governments have tended to avoid judicial and liability-based methods of dealing with [international environ-

How different things look today. With the establishment of the International Criminal Court, the International Tribunal for the Law of the Sea, and the World Trade Organization Dispute Settlement Body, the revival of arbitration under the auspices of the Permanent Court of Arbitration ("PCA"), and the emergence of many new specialized international dispute settlement bodies, concerns are now more often expressed about the proliferation of tribunals and the potential fragmentation of international law.[5] Although, even now, few international environmental disputes have been decided through traditional means,[6] cases are frequently considered by newer, quasi-legal non-compliance mechanisms.[7]

The *MOX* litigation has been a poster child for the emerging issue of fragmentation in international law. It has involved proceedings in four different forums: an arbitral tribunal convened pursuant to Article 32 of the Convention for the Protection of the Marine Environment of the North-East Atlantic ("OSPAR Convention");[8] the International Tribunal for the Law of the Sea ("ITLOS"); an arbitral panel established under Annex VII of the United Nations Convention on the Law of the Sea ("UNCLOS");[9] and the European Court of Justice. The dis-

mental questions]"). A useful compilation of early international decisions relating to the environment can be found in 1 INTERNATIONAL ENVIRONMENTAL LAW REPORTS (Cairo A.R. Robb ed., Cambridge University Press 1999). It includes a total of only 10 cases, which it describes as representing "all of the well-known and some less well-known decisions of international tribunals" during the period from 1872 to 1968.

[5] *Symposium Issue: The Proliferation of International Judicial Bodies: Piecing Together the Puzzle*, 31 NYU J. INT'L L. & POL. pp. 679–933 (1999); Jonathan I. Charney, *Is International Law Threatened by Multiple International Tribunals?*, 271 RECUEIL DES COURS p. 101 (1998). See also Colin Warbrick, *The "Iron Rhine" ("IJzeren Rijn") Arbitration (Belgium-Netherlands): Its Contribution to International Law*, THE IRON RHINE (*IJZEREN RIJN*) ARBITRATION (BELGIUM-NETHERLANDS) AWARD OF 2005 p. 1, at p. 5 (Permanent Court of Arbitration Award Series, TMC Asser Press 2007).

[6] In addition to the particular arbitration addressed in this volume, recent environmental cases include *Case Concerning the Gabčíkovo-Nagymaros Project* (Hung. v. Slovk.), 1997 I.C.J. Rep. p. 92; *Southern Bluefin Tuna* (N.Z. v. Japan, Austl. v. Japan), 38 I.L.M. p. 1624 (1999); and the *Case Concerning Pulp Mills on the River Uruguay* (Arg. v. Uru.), which is currently pending in the International Court of Justice. On international dispute settlement, see generally Cesare P.R. Romano, *International Dispute Settlement*, *in* OXFORD HANDBOOK OF INTERNATIONAL ENVIRONMENTAL LAW p. 1036 (Daniel Bodansky, Jutta Brunnee & Ellen Hey, eds., Oxford University Press 2007) [hereinafter OXFORD HANDBOOK].

[7] Jan Klabbers, *Compliance Procedures*, *in* OXFORD HANDBOOK, *supra* note 6, at p. 995.

[8] Convention for the Protection of the Marine Environment of the North-East Atlantic, 22 Sept. 1992, 32 I.L.M. p. 1069.

[9] UN Convention on the Law of the Sea, opened for signature 10 Dec. 1982, 1833 U.N.T.S. p. 397. The UNCLOS claim led to the creation of a five-member arbitral panel under Article 287

pute concerns potential radioactive pollution of the Irish Sea from a mixed oxide ("MOX") manufacturing plant at Sellafield, on the north-western coast of England. Ireland initially brought a claim against the United Kingdom under the OSPAR Convention, on the ground that the United Kingdom had failed to provide Ireland with access to relevant information regarding its decision to commission the plant. While this case was still pending, Ireland raised a separate challenge under UNCLOS, arguing that the commissioning and operation of the MOX plant violated the United Kingdom's obligations under UNCLOS to protect the marine environment. Meanwhile, the European Commission, concerned about this ongoing dispute between two European Union Member States, initiated proceedings in the European Court of Justice ("ECJ"), arguing that Ireland had violated the exclusive jurisdiction of the ECJ by submitting its dispute with the United Kingdom to arbitration under UNCLOS. Although these separate proceedings did not result in any direct conflicts – ITLOS declined to block the commissioning of the plant;[10] the *OSPAR* Arbitral Tribunal ruled against Ireland's claim to additional information; the ECJ ultimately found that it had exclusive jurisdiction over the dispute;[11] and the arbitral panel convened under UNCLOS Annex VII deferred to the ECJ's decision – the multiple proceedings raised concerns about forum shopping and jurisdictional conflict.

This volume focuses on the initial piece of the *MOX* puzzle – the proceedings brought by Ireland concerning access to information, which represented the inaugural arbitration under the OSPAR Convention. In its Final Award, the Arbitral Tribunal concluded that Article 9 of the OSPAR Convention imposes a direct duty on the United Kingdom to provide information (rather than merely a duty to establish a domestic regulatory scheme), but then construed the scope of this duty narrowly. The Award has engendered considerable commentary,[12] and is

---

and Annex VII of UNCLOS. This UNCLOS Annex VII arbitration, Ireland v. United Kingdom ("MOX Plant" Case), was also conducted under the auspices of the PCA [hereinafter *MOX Plant* Case, UNCLOS Annex VII Arbitral Tribunal]. See <www.pca-cpa.org>.

[10] *MOX Plant* Case (Ir. v. U.K.), ITLOS Case No. 10 – Request for Provisional Measures, Order of Dec. 3, 2001, 41 I.L.M. p. 405 (2002), available at <www.itlos.org> [hereinafter *MOX Plant* Case – Request for Provisional Measures].

[11] Case 459/03, *Commission of the European Communities v. Ireland*, Judgment of May 30, 2006. The ECJ found that Ireland violated Art. 292 of the Treaty Establishing the European Community, 2002 O.J. (C 325) p. 33.

[12] See Robin Churchill & Joanne Scott, *The MOX Plant Litigation: The First Half-Life*, 53 INT'L & COMP. L.Q. pp. 643–676 (2004); Malgosia Fitzmaurice, *Dispute Concerning Access to Information under Article 9 of the OSPAR Convention*, 18 INT'L J. MARINE & COASTAL LAW

notable not only for its specific conclusions regarding the nature and scope of the right of access to information, but also for its more general discussion of the relationship of the OSPAR regime to other legal rules on access to information, including European Community directives and the 1998 Aarhus Convention.[13] The majority focused closely on the obligations created by the OSPAR Convention itself, and declined to apply rules found in other regimes or that have not yet achieved the status of law. The decision's strict interpretation of the right of access to information, and its unwillingness to progressively develop the law or to look beyond the OSPAR Convention in interpreting this right, has been criticized by some commentators and environmentalists.[14] One prominent scholar, who represented Ireland in the *OSPAR* proceedings, said that the majority's approach "suggests that environmental considerations – including international legal developments which have occurred since the 1980s – have perhaps not yet fully permeated the reasoning processes of some classical international lawyers."[15] But the majority's "legalistic" approach – its focus on the treaty text rather than on broader policy considerations – should not be surprising. Ireland chose to invoke a legalized method of dispute resolution to address a comparatively narrow legal issue regarding access to information. In a field where non-legal modes of dispute resolution are on the rise, and where the lines between "hard" and "soft" law are often blurry, the Tribunal's punctilious approach to the distinction between law and policy reflected its view of its proper role as a legal institution.

II. FACTUAL AND PROCEDURAL BACKGROUND

The Sellafield nuclear facility is located in north-west England, on the Irish Sea, 184 kilometers (114 miles) from the coast of Ireland. Established in 1947 to produce plutonium and other nuclear materials for the British atomic bomb project, Sellafield was the site of the world's first civilian nuclear reactor and

---

pp. 541–558 (2003); Ted L. McDorman, *Access to Information under Article 9 of the OSPAR Convention*, 98 AM. J. INT'L L. pp. 330–339 (2004); Yuval Shany, *The First MOX Plant Award: The Need to Harmonize Competing Environmental Regimes and Dispute Settlement Procedures*, LEIDEN J. INT'L L. pp. 815–827 (2004).

[13] Convention on Access to Information, Public Participation in Decisionmaking and Access to Justice in Environmental Matters, June 28, 1998, 2161 U.N.T.S. p. 447; 38 I.L.M. p. 517 (1999).

[14] See, e.g., Shany, *supra* note 12.

[15] PHILIPPE SANDS, PRINCIPLES OF INTERNATIONAL ENVIRONMENTAL LAW pp. 857–858 (Cambridge University Press, 2d ed. 2003).

has, throughout its history, played a central role in Britain's nuclear program, both military and civilian. It covers four square kilometers, employs approximately 10,000 people, and includes plutonium production facilities, reprocessing plants, and nuclear power reactors.

Sellafield's location on the Irish Sea has long raised fears in Ireland about potential radioactive contamination and, over the years, Ireland has repeatedly sought to close the Sellafield facility. According to Ireland, nuclear discharges from Sellafield have occurred since the early 1950s, making the Irish Sea one of the most radioactively-polluted, semi-enclosed seas in the world.[16] In addition to routine discharges, the Sellafield facility has experienced several significant accidents – the most severe, the Windscale fire in 1957,[17] released large amounts of radiation into the marine environment. Although an OSPAR report noted in 2000 that the majority of radioactive inputs into the North-East Atlantic had been "drastically reduced," discharges of some radionuclides have increased since the mid-1990s as a result of the new Thermal Oxide Reprocessing Plant ("THORP") at Sellafield.[18] Radioactive discharges from Sellafield are transported by ocean currents over a wide area, and have been detected as far away as Norway, which reportedly has considered taking legal action of its own against the United Kingdom.[19]

In 1992, British Nuclear Fuels ("BNFL") – the government-owned company that operated the Sellafield facility until 2005 – decided to build a MOX manufacturing plant at Sellafield. MOX is a nuclear fuel produced from reprocessed plutonium and uranium oxides, which has been used in commercial reactors in France, Germany and Switzerland for more than twenty years. The plan for the Sellafield MOX plant was to import spent nuclear materials from foreign power plants, reprocess it at the THORP facility at Sellafield, and then use the reprocessed materials to manufacture MOX. At least initially, the MOX was to be re-exported back to foreign utilities by sea, since no nuclear reactors in the United Kingdom currently use MOX as a fuel.

---

[16] *MOX Plant* Case, UNCLOS Annex VII Arbitral Tribunal, *supra* note 9, Memorial of Ireland, July 26, 2002, p. 9, available at <www.pca-cpa.org>.

[17] Until 1981, the Sellafield facility operated under the name of Windscale.

[18] OSPAR Commission, Quality Status Report 2000 for the North East Atlantic, para. 6.6, available at <www.ospar.org>. According to the EC Commission, nuclear discharges in 1997 were only about 1% of levels in the 1970s. Churchill & Scott, *supra* note 12, at p. 673.

[19] *MOX Plant* Case, UNCLOS Annex VII Arbitral Tribunal, Memorial of Ireland, *supra* note 16, para. 1.68.

Approval of the MOX plant involved both domestic and international elements. Pursuant to domestic law, BNFL applied to the local authorities for permission to build the plant and prepared an environmental impact statement. British authorities concluded that the radioactive discharges from the MOX plant would be minimal, at levels permissible under existing authorizations.[20] Domestic approval to build the MOX plant was granted in 1994 and construction proceeded between 1994 and 1996.

Before the plant could be approved, however, the United Kingdom also needed to meet two requirements under European Community law – one concerning the plant's environmental impacts and the other concerning its economic justification – which occupied an additional five years.

First, Article 37 of the Treaty Establishing the European Atomic Energy Community ("EURATOM") requires States to provide the European Commission with information concerning plans to dispose of nuclear wastes, so that the Commission can determine whether the plan "is liable to result in the radioactive contamination of the water, soil or airspace of another Member State."[21] Pursuant to this requirement, the United Kingdom submitted detailed information to the Commission about the potential environmental impacts of the MOX plant. In February 1997, the European Commission issued an opinion that, in essence, gave the plant a clean bill of health, finding that normal discharges from the MOX plant would have "negligible" effects on other Member States, and that even unplanned discharges resulting from an accident would not likely result in "significant" radioactive contamination from "the health point of view." The Commission concluded that "both in normal operation and in the event of an accident of the type and magnitude considered in the general data," the MOX plant "is not liable to result in radioactive contamination significant from the point of view of health, of the water, soil or airspace of another Member State"[22] – findings that, according to the United Kingdom, Ireland never contested.[23]

---

[20] As the United Kingdom explained, the manufacture of MOX "is essentially a dry process so that any discharge of liquid effluent would be minimal." Ireland v. United Kingdom ("*OSPAR*" Arbitration), Counter-Memorial of United Kingdom, June 6, 2002, at para. 2.5, available at <www.pca-cpa.org>.

[21] Treaty Establishing the European Atomic Energy Community, 25 Mar. 1957, 298 U.N.T.S. p. 167, as amended by the Treaty of Nice (2001), 2001 O.J. (C 80) p. 1. The EURATOM Treaty was one of the founding instruments of the European Communities.

[22] European Commission Opinion under Article 37 of EUROTOM, Feb. 25, 1997.

[23] *OSPAR* Arbitration, Counter-Memorial of United Kingdom, *supra* note 20, at para. 2.7.

In addition to this environmental requirement, EURATOM directives require Member States to perform a cost-benefit analysis of every new activity that results in exposure to ionizing radiation, to ensure that the activity is "justified by the advantages which it produces."[24] Pursuant to these directives, the United Kingdom held five public consultations between 1997 and 2000 on the economic justification of the MOX plant, and commissioned two separate economic analyses by private consulting firms, one by the PA Consulting Group in 1997 ("PA report") and another by Arthur D. Little in 2001 ("ADL report"), both of which concluded that the plant was economically justified. On the basis of the public consultations and reports, the United Kingdom's Secretaries of State for Health and for Environment, Food and Rural Affairs determined on October 3, 2001, that the MOX plant was justified in accordance with EURATOM requirements.

The proposed MOX plant at Sellafield raised strong opposition by Ireland. Ireland wanted to see the Sellafield facility shut down, not expanded, and expressed concern about both routine discharges from the new plant and from the increased activity at the existing THORP facility, as well as about potential accidents during the transport of nuclear materials across the Irish Sea to and from Sellafield. Throughout the domestic approval process for the MOX plant, Ireland worked actively to stop it from proceeding, first by questioning the adequacy of the government's environmental impact assessment, and later, during the public consultations about the economic justification for the plant, by challenging the quality of the information provided, including the PA report, which the United Kingdom released only in redacted form. In July 1999, and again in 2000 and 2001, Ireland requested the full, unredacted version of the PA report under EC Directive 90/313 on Freedom of Access to Information on the Environment[25] and under the OSPAR Convention. The United Kingdom declined to release the full report, however, claiming that it contained confidential business information.

Faced with the imminent commissioning of the MOX plant by the United Kingdom, Ireland launched a multi-prong challenge on the basis of international

---

[24] EURATOM Directive 80/836, July 15, 1980, Art. 6. In the course of the *MOX* dispute, Directive 80/836 was replaced by EURATOM Directive 96/29, which requires that an activity's "health detriment" be justified by its "economic, social or other benefits." Directive 96/29, Art. 6(1), May 15, 1996. Directive 96/29 entered into force in May 2000 and governed the last rounds of consultations in 2001. See Final Award, para. 33.

law. Interestingly, Ireland chose not to proceed under the EURATOM Treaty (or under European Community law more generally), which establishes the most specific international regulatory regime applicable to the MOX plant, even though Ireland apparently believed that the decision to authorize the MOX plant violated several EURATOM directives.[26] Instead, it relied on two more general treaties: the OSPAR Convention, which addresses marine pollution of the North-East Atlantic, and the United Nations Convention on the Law of the Sea, which establishes a comprehensive regime for the world's oceans. Ireland initiated arbitral proceedings under the OSPAR Convention on June 15, 2001, and under Annex VII of UNCLOS on October 25, 2001.

Of the two challenges, Ireland's claims under UNCLOS were more sweeping and, in contrast to the *OSPAR* proceedings, raised environmental issues directly. Ireland alleged that the United Kingdom had violated its duties under UNCLOS to:

(1) carry out a proper environmental assessment of the likely impacts of the MOX plant on the marine environment of the Irish Sea, as required by UNCLOS Article 206;

(2) cooperate with Ireland to protect a semi-enclosed sea pursuant to UNCLOS Articles 123 and 197; and

(3) take all necessary steps to protect and preserve the marine environment of the Irish Sea pursuant to UNCLOS Articles 192, 193, 194, 207, 211, 212, 213, 217 and 222.

On November 9, 2001, Ireland applied for provisional measures to suspend authorization of the MOX plant and to prevent the marine transport of radioactive materials to and from the plant. Because there had not yet been time to constitute an arbitral panel, the Irish request for provisional measures was considered by ITLOS, which found that Ireland had failed to show that the situation was urgent and therefore declined to block the plant, although it ordered that

---

[25] Council Directive 90/313/EEC, on Freedom of Access to Information on the Environment, June 7, 1990, 1990 O.J. (L 158) p. 56.

[26] In a letter of Oct. 16, 2001, Ireland claimed that the United Kingdom's decision to authorize the MOX plant violated Directive 85/337/EEC, Directive 80/836/EURATOM, and Directive 96/239/EURATOM.

Ireland and the United Kingdom enter into consultations with one another.[27] Two years later, in June 2003, the Annex VII Arbitral Tribunal that by now had been constituted, decided to suspend further proceedings pending a resolution of the United Kingdom's objection that the European Court of Justice had exclusive jurisdiction over the issues in the case under Article 292 of the EC Treaty.[28] Later that year, the European Commission initiated an infringement proceeding against Ireland in the ECJ and, in 2006, the ECJ ruled that the claim by Ireland against the United Kingdom violated European Community law.[29]

By contrast to the UNCLOS proceeding, the *OSPAR* Arbitration involved a much narrower issue, concerning the right of access to information under Article 9 of the OSPAR Convention. Ireland claimed that the United Kingdom violated Article 9 by refusing to provide it with the unredacted version of the PA and ADL reports, which prevented Ireland from conducting a meaningful review of the economic justification for the MOX plant pursuant to EURATOM Directive 96/29. To hear Ireland's claim, a three-person arbitral panel was constituted in October 2001: two party-nominated arbitrators – Dr. Gavan Griffith QC, appointed by Ireland, and Lord Mustill, appointed by the United Kingdom – and a chairman, Professor W. Michael Reisman, selected by the party-appointed arbitrators. The Parties agreed to designate the International Bureau of the Permanent Court of Arbitration ("PCA") as the registry for the arbitration. The Arbitral Tribunal held a hearing in October 2002 and issued its Final Award on July 2, 2003.

---

[27] *MOX Plant* Case – Request for Provisional Measures, *supra* note 10. See Malcolm J.C. Forster, *The MOX Plant Case – Provisional Measures in the International Tribunal for the Law of the Sea*, 16 LEIDEN J. INT'L L. pp. 611–619 (2003).

[28] *MOX Plant* Case, UNCLOS Annex VII Arbitral Tribunal, *supra* note 9, Order No. 3, Suspension of Proceedings on Jurisdiction and Merits, June 24, 2003, 42 I.L.M. p. 1187 (2003), available at <www.pca-cpa.org>. Article 292 of the EC Treaty provides, "Member States undertake not to submit a dispute concerning the interpretation or application of this Treaty to any method of settlement other than those provided for therein."

[29] Case C-459/03, *Commission of the European Communities v. Ireland*, Judgment of May 30, 2006. See Cesare C.P. Romano, *Commission v. Ireland*, 101 AM. J. INT'L L. pp. 171–178 (2007); Nikolaos Lavranos, *MOX Plant Dispute*, 2 EUR. CONSTITUTIONAL L. REV. pp. 456–469 (2006).

III. OSPAR CONVENTION

Adopted in 1992 to combine and update two earlier treaties addressing pollution of the North-East Atlantic,[30] the OSPAR Convention consists of 34 articles and 5 annexes, which comprehensively address all sources of marine pollution of the North-East Atlantic. Among its substantive obligations, the OSPAR Convention requires parties to:

– take all possible steps to prevent and eliminate pollution (Article 2(1)(a));

– apply the precautionary principle and the polluter pays principle (Article 2(2)); and

– apply best available techniques and best environmental practice (Article 2(3)).

Specific types of marine pollution – including pollution from land-based sources, pollution by dumping, and pollution from offshore sources – are addressed in separate articles and annexes.[31] The Convention also establishes the OSPAR Commission to supervise implementation of the Convention, review the condition of the maritime area, and draw up programs and measures for the prevention and elimination of pollution (Article 10). Over the last twenty years, the OSPAR Commission (and its predecessor, the Paris Commission), at Ireland's urging, have adopted numerous measures to prevent pollution from ionizing radiation.[32]

---

[30] Oslo Convention for the Prevention of Marine Pollution by Dumping from Ships and Aircraft, Feb. 15, 1972, 932 U.N.T.S. p. 3, 11 I.L.M. p. 262; Paris Convention for the Prevention of Marine Pollution from Land-Based Sources, June 4, 1974, 1546 U.N.T.S. p. 119, 13 I.L.M. p. 352.

[31] With respect to land-based sources of marine pollution, Art. 3 of the OSPAR Convention requires parties to "take all possible steps to prevent and eliminate pollution from land-based sources." This general obligation is elaborated in Annex I (which requires parties to use best available techniques for point sources and best environmental practice for point and diffuse sources (Annex I, Art. 1.1), take preventive measures to minimize the risk of pollution caused by accidents (Art. 1.3), subject point source discharges to strict regulation (Art. 2.1), and provide for a system of regular monitoring and inspection (Art. 2.2)), as well as in various recommendations of the OSPAR Commission.

[32] See OSPAR Radioactive Substances Strategy. The decisions and recommendations of the OSPAR Commission (and its predecessor, the Paris Commission) can be found at <http://www.ospar.org/eng/html/strategies/strategy-05.htm>.

Rather than invoking these substantive environmental provisions, however, Ireland instead based its claim against the United Kingdom on the procedural requirement in Article 9 regarding access to information. Article 9 obligates the Contracting Parties "to ensure that their competent authorities are required to make available the information described in paragraph 2 of this Article to any natural or legal person, in response to any reasonable request." Paragraph 2 defines the information subject to this disclosure requirement as "any available information ... on the state of the maritime area, on activities or measures adversely affecting or likely to affect it, and on activities or measures introduced in accordance with the Convention." Paragraph 3 recognizes various exceptions, including the right of States to refuse to provide information, "in accordance with their national legal systems and applicable international regulations," where the information affects "commercial and industrial confidentiality."

Even though the OSPAR Convention is an environmental agreement, Ireland's access to information claim implicated environmental issues only obliquely, since the information in question concerned the economic justification for the MOX plant, rather than its environmental impacts. As Robin Churchill and Joanne Scott have noted, Ireland's decision to pursue its access to information claim under the OSPAR Convention rather than under European Community law was "in many ways ... surprising,"[33] since the OSPAR Convention "is not a general freedom of information statute,"[34] and the EC directive on access to information on the environment[35] "has a wider ambit" than Article 9 and has been interpreted broadly by the European Court of Justice.[36] Ireland insisted that the information it had requested, although primarily economic in orientation, was nevertheless within the scope of Article 9 because it related to an activity adversely affecting or likely to affect the marine environment.

---

[33] Churchill & Scott, *supra* note 12, at p. 674.
[34] Final Award, para. 170.
[35] *Supra* note 25.
[36] Churchill & Scott, *supra* note 12, at p. 674.

IV. THE FINAL AWARD

The case before the Arbitral Tribunal involved three issues:

(1) Whether Article 9(1) establishes a duty on the United Kingdom to directly provide information, or only a duty to establish a domestic framework for the disclosure of information?

(2) Whether the information requested by Ireland falls under Article 9(2)?

(3) If so, whether the information is privileged under Article 9(3) because of its confidential nature?

On the first issue, the two party-nominated arbitrators – Dr. Griffith and Lord Mustill (with Professor Reisman dissenting) – concluded that the OSPAR Convention imposes a direct duty on the United Kingdom to provide information, rather than simply a duty to establish a domestic regulatory framework, as the United Kingdom argued. But, on the second question, a different majority – this time consisting of Lord Mustill and Professor Reisman, with Dr. Griffith dissenting – held that the information requested by Ireland did not fall within the scope of Article 9(2), because it was neither "information ... on the state of the maritime area," nor information on activities "adversely affecting or likely to affect" the maritime area. Having concluded that the requested information was not covered by Article 9, Lord Mustill and Professor Reisman held that the Tribunal did not need to address the final question of whether the requested information is privileged under Article 9(3).

V. APPLICABLE LAW

The Arbitral Tribunal first considered the applicable law. In its pleadings, Ireland introduced materials regarding the right of access to information in other international instruments, including

– the 1998 Aarhus Convention on Access to Information, Public Participation in Decision-Making and Access to Justice in Environmental Matters –

a treaty that neither Ireland nor the United Kingdom had ratified at the time of the case and that had not yet entered into force;[37]

- two non-binding instruments, the 1992 Rio Declaration on Environment and Development[38] and the 1998 Sintra Ministerial Statement;[39] and

- various EC directives, including the directive on the freedom of access to information on the environment[40] and its successor, the directive on public access to environmental information.[41]

The majority (consisting of Professor Reisman and Lord Mustill) made four major findings regarding the applicable law.

First, the majority held that the Tribunal's subject-matter jurisdiction (jurisdiction *rationae materiae*) extended only to the Parties' obligations under the OSPAR Convention itself, not obligations created by other legal instruments, a conclusion that the dissenting arbitrator, Dr. Griffith, accepted.[42] By its terms, Article 32 of the OSPAR Convention, under which the Arbitral Tribunal operated, is limited to disputes "relating to the interpretation or application of the Convention," not other legal instruments, so the majority's conclusion on this issue is unexceptionable. Although Article 32(6)(a) directs arbitral tribunals to resolve disputes "according to the rules of international law," the majority did not think that it was "reasonable to suppose" that this language was intended to create "an unqualified and comprehensive jurisdictional regime."[43] As a result, the majority found that the Arbitral Tribunal could decide only whether the United Kingdom's refusal to provide the unredacted PA and ADL reports was consistent with its obligations under Article 9 of the OSPAR Convention, not under other legal instruments such as the Aarhus Convention or EC directives.

---

[37] The Aarhus Convention, *supra* note 13, entered into force on Oct. 30, 2001. The United Kingdom ratified it on Feb. 25, 2005, but Ireland still had not ratified the Convention as of Sept. 1, 2008.

[38] Declaration of the United Nations Conference on Environment and Development, Rio de Janeiro, June 14, 1992, 31 I.L.M. p. 874 (1992).

[39] Statement of the Ministerial Meeting of the OSPAR Commission, Sintra, Portugal, July 23, 1998.

[40] *Supra* note 25.

[41] Directive 2003/4/EC on Public Access to Environmental Information, Jan. 28, 2003, 2003 O.J. (L 41/26) p. 26.

[42] Dissenting Opinion of Gavan Griffith QC, para. 3.

[43] Final Award, para. 85.

Second, although the majority acknowledged that arbitral tribunals can look to other legal instruments in interpreting the OSPAR Convention,[44] they must do so in good faith. As the majority observed, "[a] treaty is a solemn undertaking and States Parties are entitled to have applied to them and to their peoples that to which they have agreed and not things to which they have not agreed."[45] Ultimately, the majority declined to use the definition of "environmental information" in the Aarhus Convention in interpreting the scope of Article 9, not because the use of the Aarhus Convention for interpretive purposes was impermissible, but because the language in the OSPAR Convention differs from that in Aarhus – OSPAR does not refer to "environmental information," so the definition of that term in the Aarhus Convention is not directly relevant.

Third, the Tribunal declined to apply either the Sintra Ministerial Statement (in which the OSPAR ministers agreed "to prevent pollution of the maritime area from ioning radiation through progressive and substantial reductions of discharges") or two OSPAR Commission decisions on non-reprocessing of spent nuclear fuels, which do not legally apply to the United Kingdom. In the majority's view, the Sintra Ministerial Statement was "plainly exhortatory"[46] and, in any event, does not directly concern access to information, which was the basis of Ireland's claim.

Finally, the majority insisted that, absent a specific instruction from the Parties, it could apply only existing law (*lex lata*), not "evolving" norms *in statu nascendi* – or "almost law," as the majority somewhat disparagingly characterized such norms. As the majority noted, "[a] tribunal established under the OSPAR Convention cannot go beyond existing law. This is not to say that the tribunal cannot apply customary international law of a recent vintage, but that it must in fact be customary international law."[47]

Although the majority's discussion of the applicable law has come under criticism,[48] none of these views are particularly controversial. They all reflect the commonplace view that an arbitral tribunal, as a legal body, has a limited mandate to apply existing law, rather than to develop the law in what it deems to

---

[44] Final Award, para. 105.
[45] Final Award, para. 102.
[46] Final Award, para. 91.
[47] Final Award, para. 100.
[48] E.g., Shany, *supra* note 12, at p. 822.

be a "progressive" direction. In analyzing the majority's analysis, it is important to distinguish carefully what the majority did – and did not – say. First, the majority did *not* assert that the OSPAR Convention is a self-contained regime, as some commentators have claimed.[49] In interpreting the OSPAR Convention, a tribunal can look to other legal instruments. Instead, the majority's view of the OSPAR Convention might better be characterized as that of an independent legal instrument, whose provisions have their own separate legal status and are not necessarily co-extensive with the legal rights found in other instruments. Second, the majority did not assert that interpretation is a static process, which cannot take into account developments in the law. Indeed, it agreed with Ireland that "current international law and practice" can be used in interpreting OSPAR, pursuant to Article 31(3)(b) and (c) of the Vienna Convention on the Law of Treaties.[50] The only thing a tribunal cannot do, in the majority's view, is to apply current international law and practice as such. Thus, to the extent that the OSPAR Convention establishes a narrower right of access to information than more recent practice, the tribunal must apply OSPAR, rather than the more expansive right that has developed subsequently.

VI. THE NATURE OF THE DUTY TO PROVIDE ACCESS TO INFORMATION

The first substantive issue addressed by the Tribunal concerned the nature of the duty imposed by the OSPAR Convention regarding access to information. Article 9(1) of OSPAR provides:

> The Contracting Parties shall ensure that their competent authorities are required to make available the information described in paragraph 2 of this Article in response to any reasonable request, without that person's having to prove an interest, without unreasonable charges, as soon as possible and at the latest within two months.

According to Ireland, Article 9 creates a direct obligation to provide information, which the United Kingdom violated by failing to release the full versions of the PA and ADL reports. In response, the United Kingdom argued that Article 9 imposes a duty merely to establish a domestic regulatory framework requiring disclosure of information. On this view, specific claims for access to information

[49] E.g., Shany, *supra* note 12, at p. 816.
[50] Final Award, para. 101.

(such as Ireland's request for the unredacted versions of the PA and ADL reports) are not governed by Article 9, but rather by a State's domestic regulatory framework, and should be decided by national courts rather than an international tribunal.[51] Failure to provide information in a particular case would not violate Article 9; rather, a State would violate Article 9 only if it failed to establish an appropriate regulatory framework (for example, because its framework levies "unreasonable charges" or does not require domestic authorities to provide information within two months, contrary to the terms of Article 9).

In rejecting the United Kingdom's argument, the majority found that Article 9(1) establishes an "obligation of result" – namely to make available the information specified in Article 9(2) – rather than merely an obligation to put in place a domestic regulatory framework.[52] In the majority's view, the verb used in Article 9(1) ("ensure") establishes an obligation "at the mandatory end of the scale," which must involve more than a comparatively weak obligation to establish a domestic regulatory framework.[53] The fact that Article 9 was modeled on EC Directive 90/313 (which creates only a duty to establish a domestic framework) does not mean that Article 9 is similarly limited, according to the majority, since Article 9 has "an independent legal source that establishes a distinct legal regime and provides for different legal remedies."[54] In dissent, Professor Reisman argued that the majority's conclusion violates a key canon of treaty interpretation, namely that "a legal text should be interpreted in such a way that a reason and a meaning can be attributed to every word in the text."[55] The majority's interpretation of Article 9(1) failed to do so, Professor Reisman argued, because it read Article 9(1) to mean, "the Contracting Parties shall … make available the information described in paragraph 2 of this Article," omitting the words in the ellipsis, "ensure that their competent authorities are required to."[56]

On this issue, Professor Reisman seems to have the better of the argument. The majority's analysis of the Article 9(1) issue is not altogether clear, and much

---

[51] *OSPAR* Arbitration, Counter-Memorial of United Kingdom, *supra* note 20, at paras. 3.4–3.7. The United Kingdom argued that it had satisfied the requirements of Article 9 through its adoption of the 1992 Environmental Regulations. *Id.* at para. 3.12.

[52] Final Award, para. 137.

[53] Final Award, para. 134.

[54] Final Award, para. 142.

[55] Declaration of Professor W. Michael Reisman, para. 6 (quoting the International Court of Justice in the *Anglo-Iranian Oil Company* case).

[56] Declaration of Prof. Reisman, para. 5.

of it begs the question or suggests false contrasts between its conclusion and the opposing view. For example, the Tribunal observes that Article 9(1) establishes an "enforceable obligation" rather than an "aspirational objective."[57] But reading Article 9(1) as imposing a requirement to establish a domestic regime, as the United Kingdom contended, would not make the provision merely aspirational in character. The issue in the case is not whether Article 9(1) is aspirational or mandatory; rather, the question is, what does Article 9(1) require – the actual provision of information, or merely the establishment of a domestic regulatory scheme? Similarly, the Tribunal canvasses the "carefully crafted hierarchy of obligations" in the OSPAR Convention, and notes that the verb "ensure," which is used in Article 9(1), reflects a "strong" level of expression that is at the "mandatory end of the scale."[58] Admittedly, "ensure" is a strong term, but this begs the question, what must a party ensure? The Tribunal suggests that it would be "discordant" to combine a strong term ("ensure") with a weak obligation (setting up a domestic regulatory regime). But there is nothing inconsistent or incoherent about a treaty provision obligating States to establish a domestic regulatory regime that "requires" a particular result. This is, in fact, a comparatively strong obligation, not a weak one. Finally, the majority suggests that, even if Article 9 requires States to put in place a domestic framework, States are still responsible for the adequacy of their framework. As the majority notes, "even where international law assigns competence to a national system there is no exclusion of responsibility of a State for the inadequacy of such a national system or the failure of its competent authorities to act in a way prescribed by an international obligation or implementing legislation."[59] Moreover, the OSPAR Convention clearly contemplates that domestic proceedings are not exclusive, since it creates its own international dispute settlement system under Article 32.[60] These points are all true, but there is no disagreement about any of them: the United Kingdom admitted that it could be held responsible internationally if its domestic regulatory framework was found inadequate. Its argument was simply that Ireland had not made this contention.

The Tribunal characterized the issue regarding Article 9(1) as whether the provision imposes an "obligation of result" (to make available the specified information) or an obligation of conduct (to impose a requirement on domestic

[57] Final Award, para. 127.
[58] Final Award, paras. 133–134.
[59] Final Award, para. 146.
[60] Final Award, para. 143.

authorities).[61] But this way of characterizing the issue is confusing, since the term "obligation of result" has two quite different meanings. In one sense, most international obligations could be considered "obligations of result," since, typically, international law simply requires a country to comply with its obligations, but does not specify the means it must use to do so. In this context, an obligation of result is weaker than an obligation of conduct, which specifies not simply a required result, but how a State must achieve that result – just as, in European Union law, a "directive" is a weaker type of requirement than a "regulation," since it allows States the choice of form and method. But the term "obligation of result" is also sometimes used in a quite different way. Particularly in the context of problems that result primarily from private conduct not fully controlled by the government, such as global warming, an obligation of result represents a stronger, not weaker, type of international obligation than an obligation of conduct. A country can comply with an obligation of conduct simply by performing the required action (for example, in the case of climate change, by adopting energy efficiency standards or a carbon tax), even if that conduct does not achieve the desired result (reducing carbon emissions by a particular amount). To comply with an obligation of result, in contrast, a country must do whatever it takes to achieve the required outcome.

In the *OSPAR* Arbitration, the confusion between the two senses of the term "obligation of result" is apparent from the fact that both the Tribunal and the United Kingdom argued that Article 9(1) establishes an obligation of result: the Tribunal in the sense of a stronger type of obligation, guaranteeing a particular outcome; the United Kingdom in the weaker sense of an EU directive, which leaves States with discretion regarding implementation.[62] Ironically, on either interpretation of Article 9(1), if the information requested by Ireland was objectively covered by Article 9(2) (and did not qualify for one of the exceptions in Article 9(3)), then failure by the United Kingdom to provide that information could arguably be construed as a violation of Article 9(1) – either directly (under the Tribunal's approach), or indirectly, by demonstrating that the United Kingdom's domestic regulatory scheme, as applied, did not satisfy Article 9(1), since it failed to "ensure" that domestic authorities were "required" to turn over the information specified in Article 9(2). The two interpretations end up in pretty much the same place because, even if one interprets Article 9(1) as imposing an

---

[61] Final Award, para. 137.

[62] *OSPAR* Arbitration, Counter-Memorial of United Kingdom, *supra* note 20, at para. 3.11.

obligation of conduct, the required conduct – namely, to establish a domestic regulatory framework – must ensure a particular result. The difference between the two interpretations is not that one permits an international claim by Ireland and the other consigns Ireland to UK domestic remedies; rather, the difference concerns the timing and degree of international review. Ireland contended – and the majority agreed – that it could bring an international claim directly for failure to provide covered information. In contrast, under the United Kingdom's approach, Ireland would have had to pursue the matter in British courts first, before bringing an international claim; and the British authorities would have been entitled to a "margin of appreciation" in applying their domestic framework[63] – although, even here, the term "ensure" in Article 9(1) significantly limits a State's discretion.

VII. SCOPE OF THE RIGHT OF ACCESS TO INFORMATION

Having decided that Article 9(1) establishes a direct right of access to information, the Arbitral Tribunal then turned to the question of whether the information sought by Ireland fell within the scope of Article 9(2). On this question, a different majority (this time consisting of Lord Mustill and Professor Reisman, with Dr. Griffith dissenting) held that the information was not covered by Article 9(2) and therefore the Arbitral Tribunal need not address the final question of whether the requested information is privileged under Article 9(3).

Article 9(2) encompasses three categories of information: first, "information … on the state of the maritime area"; second, information on "activities or measures adversely affecting or likely to affect" the maritime area; and third, information on "activities or measures introduced in accordance with the Convention." Lord Mustill and Professor Reisman initially concluded that the information requested by Ireland did not concern "the state of the maritime area," and therefore was not encompassed by the first category of information. As they noted, it is "manifest" that "none of the … 14 categories [of information requested by Ireland] can plausibly be characterized as 'information … on the state of the maritime area.'"[64] Moreover, Ireland had not argued that the information concerned an "activity or measure introduced in accordance with the

[63] Declaration of Prof. Reisman, at para. 18.
[64] Final Award, para. 163.

Convention," under the third category. So attention thus focused on the second category, namely, information on activities or measures "adversely affecting or likely to affect" the maritime area.

Ireland contended that this category should be given a broad scope, encompassing any information relevant to the decision to approve an activity affecting the marine environment. In support of this view, it argued that other international instruments interpret the right of access to information broadly, including EC Directives 90/313 and 2003/4, the 1998 Aarhus Convention, and Principle 10 of the 1992 Rio Declaration. The 1998 Aarhus Convention, in particular, explicitly encompasses within its definition of "environmental information" exactly the type of information at issue in the *OSPAR* Arbitration, namely, information on "cost-benefit and other economic analyses and assumptions used in environmental decision-making."[65] In Ireland's view, Article 9 of the OSPAR Convention should be interpreted in light of this emerging right to information under international environmental law.

The majority's rejection of Ireland's claim has been read by some as resting on a narrow interpretation of what constitutes "environmental information," which ignores these recent developments in international environmental law. In its pleadings, the United Kingdom characterized the Irish position as depending on a principle of "inclusive causality," requiring disclosure of information about every step in the causal chain leading to decisions about activities that might affect the marine environment. Instead, the United Kingdom argued that Article 9 should be read more narrowly, as limited to information about the environmental effects of an activity. At times, the majority's opinion seems to endorse this view about the limited types of information covered by Article 9(2) and to reject the principle of inclusive causality. To the extent this is true, then the majority's conclusion runs counter to its observation that the reference to "any information" in Article 9(2) "requires an applier to interpret extensively *within* each of the three categories" and not to imply any selection or restriction.[66] As the majority notes, "[o]nce a matter is found to fall within one of the categories of Article 9(2), the presumption is that it is within the scope of the OSPAR Convention." Under this reasoning, economic information on an activity likely to affect the maritime area adversely is not *ipso facto* outside the scope of Article 9(2), even if it does not

---

[65] Aarhus Convention, *supra* note 13, para. 2.3(b).
[66] Final Award, para. 167.

constitute "environmental information," since Article 9(2) is not limited to environmental information but instead encompasses "any information." To the extent that the majority's opinion suggests otherwise, and rejects the theory of inclusive causality, then this conclusion is inconsistent with the majority's own analysis of Article 9(2).

The majority also puts forward an alternative argument, however, which has a more solid basis, namely that the information requested by Ireland was outside the scope of Article 9(2), not because it focused on economic rather than environmental information, but because it did not meet the threshold requirement that it related to an activity "adversely affecting or likely to affect" the marine environment. In its claim, Ireland had not focused on the environmental impact of the MOX plant, instead emphasizing its right of access to information. Dr. Griffith argued in dissent that Ireland did not need to establish the risk of environmental harm, since according to the precautionary principle, the burden of proof should be on the United Kingdom to show that the MOX plant was safe, not on Ireland to show that it was hazardous.[67] But, although the OSPAR Convention, in general, endorses the precautionary principle, the specific language of Article 9(2) takes a narrower approach, requiring a "likelihood" of adverse impacts. Moreover, although the majority put forward no evidence in support of its conclusion that the MOX plant was not "likely" to harm the marine environment,[68] the European Commission had previously concluded that the plant would have a "negligible" effect – a conclusion that Ireland apparently did not specifically challenge.[69]

Of course, if Ireland had been seeking information about the environmental impacts of the MOX plant, then it might have had a stronger argument that it could not be expected to establish a likelihood of adverse environmental effects in order to have a right to the information, when that was the very issue about which it was seeking information in the first place. But in this case, Ireland was not seeking information about environmental impacts; rather it was seeking information on the economic justification for the plant. So the United Kingdom's refusal of Ireland's request did not prejudice the ability of Ireland to show that the MOX plant was likely to have adverse environmental effects, in order to establish its claim under Article 9(2).

---

[67] Dissenting Opinion of Gavan Griffith QC, paras. 72–73.

[68] Dissenting Opinion of Gavan Griffith QC, para. 92.

[69] See text accompanying notes 22 & 23, *supra*.

VIII. CONCLUSION

In many contexts, the distinction between "hard" and "soft" law often seems more apparent than real. On the one hand, "hard" norms often go unenforced and have relatively little effect on conduct; on the other, "soft" norms can be hugely influential.[70] But one arena in which the distinction between hard and soft law remains crucial is legal dispute resolution. After reading Article 9(1) of the OSPAR Convention rather loosely, the majority interpreted Article 9(2) strictly, refusing to interpret its scope in light of emerging norms that had not yet achieved the status of law. In doing so, the majority not only decided the particular issues in the case, but reaffirmed the distinctive character of legalized methods of dispute settlement.

---

[70] *See generally* DINAH SHELTON, COMMITMENT AND COMPLIANCE: THE ROLE OF NON-BINDING NORMS IN THE INTERNATIONAL LEGAL SYSTEM (Oxford University Press 2003); Kenneth Abbott & Duncan Snidal, *Hard and Soft Law in International Governance*, 54(3) INT'L ORGANIZATION pp. 421–456 (2000).

PERMANENT COURT OF ARBITRATION

DISPUTE CONCERNING ACCESS TO INFORMATION UNDER ARTICLE 9 OF THE OSPAR CONVENTION

IRELAND

VERSUS

UNITED KINGDOM OF GREAT BRITAIN AND NORTHERN IRELAND

FINAL AWARD

TRIBUNAL:

Professor W. Michael Reisman, Chairman

Dr. Gavan Griffith QC

Lord Mustill

The Hague, 2 July 2003

Table of Contents

I. INTRODUCTION

1. This matter concerns a dispute between the Ireland as claimant and the United Kingdom of Great Britain and Northern Ireland ("the United Kingdom") as respondent, determined by a Tribunal constituted pursuant to the 1992 Convention for the Protection of the Marine Environment of the North-East Atlantic ("the OSPAR Convention").[1] The issue concerns access to information as defined by the OSPAR Convention. Ireland has requested access to information redacted from reports prepared as part of the approval process for the commissioning of a Mixed Oxide Plant ("the MOX Plant") in the United Kingdom, based on Ireland's understanding of Article 9 of the OSPAR Convention. The United Kingdom has declined to provide the information requested based on its understanding of the OSPAR Convention.

II. THE OSPAR CONVENTION

2. The OSPAR Convention comprises 34 articles, five annexes and three appendices. Under Article 14(1), "[t]he Annexes and Appendices form an integral part of the OSPAR Convention." Article 14(2) provides: "The Appendices shall be of a scientific, technical or administrative nature."

3. Article 1 sets out definitions, to be considered as necessary in the course of this award. Article 2 provides:

> **GENERAL OBLIGATIONS**
>
> 1. (a) The Contracting Parties shall, in accordance with the provisions of the Convention, take all possible steps to prevent and eliminate pollution and shall take the necessary measures to protect the maritime area against the adverse effects of human activities so as to safeguard human health and to conserve marine ecosystems and, when practicable, restore marine areas which have been adversely affected.
>
>   (b) To this end Contracting Parties shall, individually and jointly, adopt programmes and measures and shall harmonise their policies and strategies.
>
> 2. The Contracting Parties shall apply:
>
>   (a) the precautionary principle, by virtue of which preventive measures are to be taken when there are reasonable

1. Convention for the Protection of the Marine Environment of the North-East Atlantic, 22 September 1992, 32 ILM 1069 (1992). Ireland and the United Kingdom are both Parties to the OSPAR Convention.

grounds for concern that substances or energy introduced, directly or indirectly, into the marine environment may bring about hazards to human health, harm living resources and marine ecosystems, damage amenities or interfere with other legitimate uses of the sea, even when there is no conclusive evidence of a causal relationship between the inputs and the effects;

(b) the polluter pays principle, by virtue of which the costs of pollution prevention, control and reduction measures are to be borne by the polluter.

3. (a) In implementing the Convention, Contracting Parties shall adopt programmes and measures which contain, where appropriate, time limits for their completion and which take full account of the use of the latest technological developments and practices designed to prevent and eliminate pollution fully.

(b) To this end they shall:

(i) taking into account the criteria set forth in Appendix 1, define with respect to programmes and measures the application of, *inter alia*,

– best available techniques

– best environmental practice

including, where appropriate, clean technology;

(ii) in carrying out such programmes and measures, ensure the application of best available techniques and best environmental practice as so defined, including, where appropriate, clean technology.

4. The Contracting Parties shall apply the measures they adopt in such a way as to prevent an increase in pollution of the sea outside the maritime area or in other parts of the environment.

5. No provision of the Convention shall be interpreted as preventing the Contracting Parties from taking, individually or jointly, more stringent measures with respect to the prevention and elimination of pollution of the maritime area or with respect to the protection of the maritime area against the adverse effects of human activities.

4. Article 3 provides:

POLLUTION FROM LAND-BASED SOURCES

The Contracting Parties shall take, individually and jointly, all possible steps to prevent and eliminate pollution from land-based sources in accordance with the provisions of the Convention, in particular as provided for in Annex I.

5. Article 4 provides:

POLLUTION BY DUMPING OR INCINERATION

The Contracting Parties shall take, individually and jointly, all possible steps to prevent and eliminate pollution by dumping or incineration of wastes or other matter in accordance with the provisions of the Convention, in particular as provided for in Annex II.

6. Article 9 provides:

ACCESS TO INFORMATION

1. The Contracting Parties shall ensure that their competent authorities are required to make available the information described in paragraph 2 of this Article to any natural or legal person, in response to any reasonable request, without that person's having to prove an interest, without unreasonable charges, as soon as possible and at the latest within two months.

2. The information referred to in paragraph 1 of this Article is any available information in written, visual, aural or data base form on the state of the maritime area, on activities or measures adversely affecting or likely to affect it and on activities or measures introduced in accordance with the Convention.

3. The provisions of this Article shall not affect the right of Contracting Parties, in accordance with their national legal systems and applicable international regulations, to provide for a request for such information to be refused where it affects:

   (a) the confidentiality of the proceedings of public authorities, international relations and national defence;

   (b) public security;

   (c) matters which are, or have been, *sub judice*, or under enquiry (including disciplinary enquiries), or which are the subject of preliminary investigation proceedings;

(d) commercial and industrial confidentiality, including intellectual property;

(e) the confidentiality of personal data and/or files;

(f) material supplied by a third party without that party being under a legal obligation to do so;

(g) material, the disclosure of which would make it more likely that the environment to which such material related would be damaged.

4. The reasons for a refusal to provide the information requested must be given.

7. Article 10 establishes a Commission of representatives of each of the Contracting Parties, sets out its duties, and, with respect to those duties, authorizes the Commission to "*inter alia*, adopt decisions and recommendations in accordance with Article 13." Article 13(1) states that "[d]ecisions and recommendations shall be adopted by unanimous vote of the Contracting Parties." If unanimity is not attainable, decisions may be taken by a three-quarters majority vote of the Contracting Parties and will become binding on those voting for it, if, at the end of 200 days after its adoption the number of Contracting Parties who have notified the Executive Secretary that they are unable to accept the decision does not reduce the number of those accepting the decision to below three-quarters of the Contracting Parties to the OSPAR Convention.

8. The provisions for amendment of the OSPAR Convention, addition and amendment of annexes, and addition and amendment of appendices are not relevant to the issues in dispute in this case.

9. Article 28 provides that no reservations may be made to the Convention.

10. Article 31 provides for the continuing force of decisions, recommendations, and all other agreements adopted under the Oslo and Paris Conventions[2] to the extent that they are compatible with the OSPAR Convention and have not been terminated by its procedures.

2. The Convention for the Prevention of Marine Pollution by Dumping from Ships and Aircraft (Oslo), 932 UNTS 3 (1972), and the Convention for the Prevention of Marine Pollution from Land-based Sources (Paris), 13 ILM 352 (1974).

11. Article 32 of the OSPAR Convention provides:[3]

SETTLEMENT OF DISPUTES

1. Any disputes between Contracting Parties relating to the interpretation or application of the Convention, which cannot be settled otherwise by the Contracting Parties concerned, for instance by means of inquiry or conciliation within the Commission, shall at the request of any of those Contracting Parties, be submitted to arbitration under the conditions laid down in this Article.

2. Unless the Parties to the dispute decide otherwise, the procedure of the arbitration referred to in paragraph 1 of this Article shall be in accordance with paragraphs 3 to 10 of this Article.

3. (a) At the request addressed by one Contracting Party to another Contracting Party in accordance with paragraph 1 of this Article, an arbitral tribunal shall be constituted. The request for arbitration shall state the subject matter of the application including in particular the Articles of the Convention, the interpretation or application of which is in dispute.

(b) The applicant party shall inform the Commission that it has requested the setting up of an arbitral tribunal, stating the name of the other party to the dispute and the Articles of the Convention the interpretation or application of which, in its opinion, is in dispute. The Commission shall forward the information thus received to all Contracting parties to the Convention.

4. The arbitral tribunal shall consist of three members: each of the parties to the dispute shall appoint an arbitrator; the two arbitrators so appointed shall designate by common agreement the third arbitrator who shall be the chairman of the tribunal. The latter shall not be a national of one of the parties to the dispute, nor have his usual place of residence in the territory of one of these parties,

---

3. It should also be noted that the last preambular paragraph of the Rules of Procedure for this Tribunal (the "Rules of Procedure for the Tribunal Constituted Under the OSPAR Convention Pursuant to the Request of Ireland dated 15th June 2001") provides:

> Whereas the Applicant and the Respondent (together, the 'Parties') have decided that the procedure of the arbitration of the Dispute shall be in accordance with the following rules (the 'Rules'), which shall replace Articles 32(4) to 32(10) of the OSPAR Convention, insofar as they do not impair the rights of other States Parties to the OSPAR Convention.

The OSPAR Tribunal Rules of Procedure may be found on the website of the Permanent Court of Arbitration (hereinafter "PCA") at www.pca-cpa.org.

nor be employed by any of them, nor have dealt with the case in any other capacity.

5. (a) If the chairman of the arbitral tribunal has not been designated within two months of the appointment of the second arbitrator, the President of the International Court of Justice shall, at the request of either party, designate him within a further two months' period.

(b) If one of the parties to the dispute does not appoint an arbitrator within two months of receipt of the request, the other party may inform the President of the International Court of Justice who shall designate the chairman of the arbitral tribunal within a further two months' period. Upon designation, the chairman of the arbitral tribunal shall request the party which has not appointed an arbitrator to do so within two months. After such period, he shall inform the President of the International Court of Justice who shall make this appointment within a further two months' period.

6. (a) The arbitral tribunal shall decide according to the rules of international law and, in particular, those of the Convention.

(b) Any arbitral tribunal constituted under the provisions of this Article shall draw up its own rules of procedure.

(c) In the event of a dispute as to whether the arbitral tribunal has jurisdiction, the matter shall be decided by the decision of the arbitral tribunal.

7. (a) The decisions of the arbitral tribunal, both on procedure and on substance, shall be taken by majority voting of its members.

(b) The arbitral tribunal may take all appropriate measures in order to establish the facts. It may, at the request of one of the parties, recommend essential interim measures of protection.

(c) If two or more arbitral tribunals constituted under the provisions of this Article are seized of requests with identical or similar subjects, they may inform themselves of the procedures for establishing the facts and take them into account as far as possible.

(d) The parties to the dispute shall provide all facilities necessary for the effective conduct of the proceedings.

(e) The absence or default of a party to the dispute shall not constitute an impediment to the proceedings.

8. Unless the arbitral tribunal determines otherwise because of the particular circumstances of the case, the expenses of the tribunal, including the remuneration of its members, shall be borne by the parties to the dispute in equal shares. The tribunal shall keep a record of all its expenses, and shall furnish a final statement thereof to the parties.

9. Any Contracting Party that has an interest of a legal nature in the subject matter of the dispute which may be affected by the decision in the case, may intervene in the proceedings with the consent of the tribunal.

10. (a) The award of the arbitral tribunal shall be accompanied by a statement of reasons. It shall be final and binding upon the parties to the dispute.

(b) Any dispute which may arise between the parties concerning the interpretation or execution of the award may be submitted by either party to the arbitral tribunal which made the award or, if the latter cannot be seized thereof, to another arbitral tribunal constituted for this purpose in the same manner as the first.

12. Annex I deals with the prevention and elimination of pollution from land-based sources. Article 2 of this annex establishes obligations with respect to them.

13. Article 3 of Annex II provides:

1. The dumping of all wastes or other matter is prohibited, except for those wastes or other matter listed in paragraphs 2 and 3 of this Article.

2. The list referred to in paragraph 1 of this Article is as follows:

(a) dredged material;

(b) inert materials of natural origin, that is solid, chemically unprocessed geological material the chemical constituents of which are unlikely to be released into the marine environment;

(c) sewage sludge until 31st December 1998;

(d) fish waste from industrial fish processing operations;

(e) vessels or aircraft until, at the latest, 31st December 2004.

3. (a) The dumping of low and intermediate level radioactive substances, including wastes, is prohibited.

(b) As an exception to subparagraph 3(a) of this Article, those Contracting Parties, the United Kingdom and France, who wish to retain the option of an exception to subparagraph 3(a) in any case not before the expiry of a period of 15 years from 1st January 1993, shall report to the meeting of the Commission at Ministerial level in 1997 on the steps taken to explore alternative land-based options.

(c) Unless, at or before the expiry of this period of 15 years, the Commission decides by a unanimous vote not to continue the exception provided in subparagraph 3(b), it shall take a decision pursuant to Article 13 of the Convention on the prolongation for a period of 10 years after 1st January 2008 of the prohibition, after which another meeting of the Commission at Ministerial level shall be held. Those Contracting Parties mentioned in subparagraph 3(b) of this Article still wishing to retain the option mentioned in subparagraph 3(b) shall report to the Commission meetings to be held at Ministerial level at two yearly intervals from 1999 onwards about the progress in establishing alternative land-based options and on the results of scientific studies which show that any potential dumping operations would not result in hazards to human health, harm to living resources or marine ecosystems, damage to amenities or interference with other legitimate uses of the sea.

14. The United Kingdom's signature was accompanied by the following declaration:[4]

> The Government of the United Kingdom of Great Britain and Northern Ireland declares its understanding of the effect of paragraph 3 of Article 3 of Annex II to the Convention to be amongst other things that, where the Commission takes a decision pursuant to Article 13 of the Convention, on the prolongation of the prohibition set out in subparagraph (3)(a), those Contracting Parties who wish to retain the option of the exception to that prohibition as provided for in subparagraph (3)(b) may retain that option,

---

4. The declaration may be found at www.ospar.org, where a note from the OSPAR Secretariat follows the declaration:

> Following the entry into force of OSPAR Decision 98/2 on Dumping of Radioactive Waste on 9 February 1999, subparagraphs (b) and (c) of paragraph 3 of Article 3 of Annex II to the Convention ceased to have effect.

provided that they are not bound, under paragraph 2 of Article 13, by that decision.

III. FACTUAL BACKGROUND

15. British Nuclear Fuels, plc ("BNFL"), a public limited company wholly owned by the United Kingdom, owns and operates a licensed nuclear enterprise at Sellafield in Cumbria. In 1993, BNFL applied to the local authority for permission to build a MOX Plant to process spent nuclear fuels by retrieving and blending separated plutonium oxide and uranium oxide into pellets to be reused as fuel in nuclear reactors. BNFL prepared and submitted Environmental Statements to the relevant authorities,[5] as required by United Kingdom law.[6] Relevant consents to build the Plant were given in 1994, and construction was completed in 1996.

16. Each of Ireland and the United Kingdom is a Party to the Treaty Establishing the European Atomic Energy Community ("EURATOM"),[7] which includes a comprehensive regulatory system for planning for the disposal of radioactive waste. Article 37 of EURATOM provides:

> Each Member State shall provide the Commission with such general data relating to any plan for the disposal of radioactive waste in whatever form as will make it possible to determine whether the implementation of such plan is liable to result in the radioactive contamination of the water, soil or airspace of another Member State.
>
> The Commission shall deliver its opinion within six months, after consulting the group of experts referred to in Article 31.

In *Saarland v. Minister for Industry*, the European Court explained the purpose of the Article 37 procedure as follows:

> [t]he purpose of Article 37, within the context of environmental protection, is to provide the Commission with comprehensive information on every plan for disposal and every activity liable to cause accidental discharges of

---

5. Ireland's Memorial, Annex 9. The Parties' written pleadings are available at www.pca-cpa.org. Annexes are on file at the offices of the PCA.

6. UK Town and Country Planning (Assessment of Environmental Effects) Regulations 1988 (SI No. 1199).

7. Treaty Establishing the European Atomic Energy Community ("EURATOM"), 25 March 1957, 298 UNTS 167.

waste, so that it is in a position to assess the repercussions thereof on the environment in the other Member States.[8]

17. On 2 August 1996, the United Kingdom submitted the data required under Article 37 to the European Commission. On 25 February 1997, the European Commission delivered its opinion under Article 37, including the conclusions:

(a) The distance between the plant and nearest point on the territory of another Member State, in this case Ireland, is 184 km;

(b) Under normal operating conditions, the discharge of liquid and gaseous effluents will be small fractions of present authorized limits and will produce an exposure of the population in other Member States that is negligible from the health point of view;

(c) Low-level solid radioactive waste is to be disposed to the authorized Drigg site operated by BNFL plc. Intermediate level wastes are to be stored at the Sellafield site, pending disposal to an appropriate authorized facility;

(d) In the event of unplanned discharges of radioactive waste which may follow an accident on the scale considered in the general data, the doses likely to be received by the population in other Member States would not be significant from the health point of view.

In conclusion, the Commission is of the view that the implementation of the plan for the disposal of radioactive wastes arising from the operation of the BNFL Sellafield mixed oxide fuel plant, both in normal operation and in the event of an accident of the type and magnitude considered in the general data, is not liable to result in radioactive contamination, significant from the point of view of health, of the water, soil or airspace of another Member State.[9]

18. Although the Article 37 procedure fulfilled a critical part of the United Kingdom's international legal obligations with respect to the environmental consequences of commissioning, there were further requirements under EURATOM and United Kingdom law to be met before the MOX Plant could be commissioned and operated. Relevantly, the domestic agency approving the Plant was required to ensure whatever environmental detriments it might cause were

---

8. Case 187/87, *Saarland and Others v. Minister for Industry, Post and Telecommunications and Tourism and Others* (reference for a preliminary ruling from the *tribunal administratif*, Strasbourg), [1988] ECR 5013, at p. 5018.

9. European Commission Opinion under Article 37 EURATOM, 1997 OJ (C 86) 3. *See* United Kingdom's Counter-Memorial, Annex 9 (Vol. II).

economically justified. In its most recent formulation, Directive 96/29 EURATOM provided in Article 6(1) that:

> Member States shall ensure that all new classes or types of practice resulting in exposure to ionizing radiation are justified in advance of being first adopted or first approved by their economic, social or other benefits in relation to the health detriment they may cause.[10]

19. Although the relevant United Kingdom statute, the United Kingdom Radioactive Substances Act 1993, does not, in terms, require such justification, in *R v. Secretary of State for the Environment and others ex parte Greenpeace Ltd.*,[11] Potts J. held (as explained in a later case) that "there was a legal obligation to justify any activity resulting in exposure to ionizing radiation in accordance with the then operative Directive, namely Euratom 80/836."[12]

20. Accordingly, over a period of eight weeks in 1997, the United Kingdom Environment Agency ("the Agency") held a public consultation on the economic justification of the MOX Plant at Sellafield.[13] This initial public consultation emerged as the first of five such consultations.

21. By its letter of 5 February 1997 inviting views, the Agency stated:

> The Agency considers that the issues associated with uranium commissioning may be separated from those associated with full operation and are simpler in nature, since no plutonium is involved. BNFL has also stated that the total activity discharged would be very small, amounting to less than 0.0000001% of the total activity discharged from the Sellafield site.[14]

The Agency enclosed a document entitled "Radioactive Substances Act 1993 Explanatory Memorandum for the Consultation on Justification and Uranium

---

10. Directive 96/29 EURATOM, Article 6(1), 1996 OJ (L 159) 1. Several documents cited below refer instead to an earlier version of this directive – namely, Directive 80/836 EURATOM, 1980 OJ (L 246) 1.

11. *R v. Secretary of State for the Environment and others ex parte Greenpeace Ltd.*, [1994] 4 All ER 352.

12. *R (Friends of the Earth Ltd. and Greenpeace Ltd.) v. Secretary of State for the Environment, Food and Rural Affairs and Secretary of State for Health*, [2001] EWHC Admin. 914, at para. 8.

13. As argued by Counsel for the United Kingdom, Oral Hearing Transcript (hereinafter "Transcript"), Day 2 Proceedings, pp. 64-66. Transcripts are available at www.pca-cpa.org.

14. Letter from I.T. Porter, Environment Agency to statutory consultees (February 5, 1997), at Tab 1, p. 2, *in* SMP Consultation Documents Bundle ("SMP Bundle"), on file at the offices of the PCA.

Commissioning of Sellafield MOX Plant (SMP)," which explained in its introduction that:

> [a]ll practices giving rise to radioactive waste must be justified, i.e. the benefits of the practice must outweigh the detriments. The manufacture of fuel in the Sellafield MOX Plant (SMP) is a new practice on the Sellafield site. The need for the Agency to consider justification in advance of the commissioning and operation of SMP arises from EU Council Directive of 15 July 1980, which lays down the basic safety standards for the health protection of the general public and workers against dangers of ionizing radiation (the Euratom Directive) . . . .[15]

The Agency went on to explain that those issues did not have to be part of the application because no change in the estimated radiological impact of the predicted operational releases was anticipated and no change in permitted levels was being requested. Rather, the focus would be on economic justification.[16] Nonetheless, data on projected aerial and liquid discharges was included.[17]

22. Another enclosure with the 5 February letter was an undated document entitled "Sellafield MOX Plant (SMP)," which had been transmitted to the Agency by BNFL under a covering letter of 27 January 1997. The transmittal letter identified the document as "the public consultation document" covering commercial and dose aspects. In discussing waste management, the document referred to "effluent arisings" which would "be conditioned as necessary to make them suitable, after monitoring, for discharge to sea."[18]

23. The Government of Ireland participated in this first of the public consultations as a "respondent." In its submission dated 4 April 1997, Ireland stated that it "opposes the commissioning of the MOX Plant on the grounds that it will perpetuate the nuclear fuel reprocessing industry in Britain," and that it deemed "objectionable and unacceptable" the "additional radioactive marine discharges from Sellafield into the Irish Sea arising from MOX production."[19] Ireland went on to raise several specific concerns about the proposed MOX Plant, including

15. "UK Environment Agency, Radioactive Substances Act 1993, Explanatory Memorandum for the Consultation on Justification and Uranium Commissioning of Sellafield Mox Plant (SMP), British Nuclear Fuels plc at Sellafield", *in* SMP Bundle, at Tab 1, p. 3, para. 1.3 (1997).

16. *Id.*, at p. 4, para. 1.6.

17. *Id.*, at p. 11, para. 6.2.

18. "Sellafield MOX Plant (SMP)", attachment to Letter from Robert Anderson of BNFL to the UK Environment Agency (January 27, 1997), *in* SMP Bundle, at Tab 1, p. 13.

19. Ireland's Memorial, Annex 4, at No. 2.

one that "the quality of information available for consultation is deficient in many respects."[20]

24. The initial consultations were followed by a further round of public consultations because "several respondents . . . were concerned that BNFL had not provided in the public domain sufficient commercial information to justify the commissioning and operation of the plant."[21] Further, other respondents had raised concerns about whether "the movements of MOX fuel from the SMP by air, sea or land"[22] could be carried out safely, and still other respondents raised non-proliferation and other security concerns.[23]

25. In preparation for the second consultation, the Agency asked BNFL to provide additional information in the form of a business case that could be independently examined. It invited prominent financial consultants to tender for the work and selected the PA Consulting Group, London ("PA") to carry out a detailed assessment.[24] As the Agency's Explanatory Memorandum under the Radioactive Substances Act explained, in addition, "PA was requested to identify if there were areas of the economic case that were not commercially sensitive which could be published in the public domain."[25]

26. PA submitted the full version of its report ("the PA Report") to BNFL and, pursuant to the Agency's request, then considered what data should be redacted. After consulting with BNFL about redactions, PA made recommendations, which were reviewed and finally determined by the Agency, and reflected in a public version of the PA Report released in December 1997 ("the 1997 PA Report"). PA gave a detailed explanation of the basis for redactions from its full report on "commercial confidentiality" grounds under section 4(2) of the United

20. *Ibid.*

21. Letter from UK Environment Agency to Friends of the Earth (January 14, 1998), *in* SMP Bundle, at Tab 2.

22. "UK Environment Agency, Explanatory Memorandum for a Further Public Consultation on the Application by BNFL for the Commissioning and Operation of the Mixed Oxide Fuel Plant at its Sellafield site in Cumbria", *in* SMP Bundle, at Tab 2, p. 4, para. 2.2 (1998).

23. *Id.*, at p. 4, para. 2.3.

24. *Id.*, at pp. 1-2.

25. *Id.*, at p. 2.

Kingdom's Environmental Information Regulations (1992) ("the 1992 Regulations"),[26] and stated:

> 1.3. COMMERCIAL CONFIDENTIALITY ISSUES
>
> PA was asked to provide the Agency with an independent view on the validity of BNFL's assertion that elements of the economic case for the SMP are commercially sensitive, and therefore that certain information therein should be withheld from the public domain. The Environmental Information Regulations 1992 (section 4(2)) provide that for the purposes of those regulations 'information relating to matters to which any commercial or industrial confidentiality attaches' may be treated as confidential. PA therefore identified a series of specific criteria to determine the information the placing of which in the public domain could prejudice the commercial interests of BNFL. Information should not be placed in the public domain if it would:
>
> 1. Allow or assist competitors to build market share or to benchmark their own operations.
>
> 2. Allow or assist competitors to attack the BNFL customer base and erode business profitability.
>
> 3. Allow or assist new competitors to enter the market.
>
> 4. Allow customers or competitors to understand the specific economics and processes of the BNFL MOX fuel fabrication business.
>
> 5. Breach contractual confidentiality requirements with customers or vendors.
>
> In addition, information should not be placed in the public domain that would breach security and safeguards requirements with respect to plutonium quantities, locations and movements.
>
> Given the issue of commercial confidentiality, and using the criteria set out above, two parallel reports have been prepared. The version for the Environment Agency contains information commercially confidential to BNFL; in the public domain report PA has replaced this information with a box in which is outlined the nature of the confidential information that has been removed and the reason, in terms of the criteria set out above, for the removal. In addition, in certain instances, specific financial, production or customer data in the full report have been deleted or replaced by a word such as 'significant' or 'minor' in the public domain version. These represent the only differences between the texts of the full version and the

26. UK Environmental Information Regulations 1992 (SI No. 1992/3240).

> public domain version. This approach enables the placing in the public domain of information that allows public review of the robustness of the BNFL economic case, without prejudicing the commercial interests of BNFL.[27]

27. In the second public consultation, Ireland submitted a detailed statement which was critical of parts of the reasoning of the 1997 PA Report. Although it did not then object to or mention any of the redactions from the published version of the PA Report, Ireland's submission concluded:

> [T]he PA Consulting report has failed to fulfil the purpose of this further consultation as set out in the Environment Agency's letter of 14 January, 1998, namely, to provide in the public domain sufficient commercial information to justify the commissioning and operation of the plant.[28]

28. After consultations, in October 1998 the Agency released a draft decision which found that "[t]he assessed radiation doses to members of the public as a consequence of discharges from the MOX Plant have negligible radiological significance,"[29] and that the balance between benefits and detriments was "broadly neutral" in terms of radioactive discharges, waste management, health and safety operations on the MOX Plant's transport, safety of the MOX fuel in reactors, radiological impact, sustainable development and proliferation of nuclear weapons and the plutonium stockpile.[30] The decision then considered the question of economic justification, which it found compelling.

29. Among others, Ireland made further representations,[31] whereafter a decision was taken in June 1999 at the ministerial level to release a new version of the PA Report, with some of the redacted material restored, and to hold a third round of consultations. The Department for the Environment, Transport and the Regions ("DETR") and the Ministry of Agriculture, Fisheries and Food ("MAFF") invited

27. "PA Consulting Group, Environment Agency Final Report – Public Domain Version; Assessment of BNFL's Economic Case for the Sellafield MOX Plant", *in* SMP Bundle, at Tab 2, pp. 1-1 to 1-6 (1997).

28. Ireland's Memorial, Annex 4, at No. 3.

29. "UK Environment Agency, Radioactive Substances Act 1993, Document Containing the Agency's Proposed Decision on the Justification for the Plutonium Commissioning and Full Operation of the MOX Plant, BNFL plc. at Sellafield", *in* SMP Bundle, Tab 2, at para. 22 (1998).

30. *Id.*, at para. 31.

31. Counsel for the United Kingdom, Transcript, Day 2 Proceedings, p. 69.

fresh comments "on the material concerning the economic case for the [MOX] plant."[32]

30. Ireland again submitted comments. By letter of 30 July 1999, Ireland elaborated upon its earlier objections and requested "an unedited and full copy of the PA Report."[33] Among other things, Ireland argued that "the information made available in the June 1999 version of the PA Report does not provide a basis for concluding that the MOX Plant is 'justified' within the meaning of Directive 80/836 EURATOM (as amended)." Ireland also raised the issue of compliance with EC Directive 90/313/EEC ("Directive 90/313")[34] on Freedom of Access to Environmental Information, and reserved its right

> to invoke – *inter alia* in relation to intensified international transportation associated with the MOX plant – procedures and substantive requirements under *inter alia* . . . the 1992 OSPAR Convention.

31. The process of review was interrupted in 1999 when BNFL discovered that fuel pellet diameter readings at the MOX demonstration facility had been falsified and reported this fact to the nuclear installation inspectorate.

32. The OSPAR Convention was first raised by Ireland in connection with the redacted information in the PA Report on 25 May 2000, when it wrote to DETR, invoking Article 9 in requesting information redacted from the published PA Report.[35] On 27 October 2000, DETR responded that "the UK Government does not wish to prejudice the commercial interests of an enterprise by disclosing commercially confidential information."[36]

33. In March 2001, a fourth consultation process commenced, now under Directive 96/29 EURATOM (*see* para.18 above), which had come into force in May 2000. In the consultation paper issued by DETR and the Department of Health in March 2001, potential respondents were invited to comment on BNFL's economic case,

---

32. Letter from UK DETR & MAFF (June 11, 1999), *in* SMP Bundle, at Tab 3.

33. Ireland's Memorial, Annex 4, at No. 4.

34. Directive 90/313/EEC, 1990 OJ (L 158) 56.

35. Ireland's Memorial, Annex 4, at No. 9.

36. *Id.*, at No. 10.

as revised in light of the data falsification incident, and an updated MOX market review.[37]

34. Ireland filed a detailed submission dated 22 May 2001. Ireland concluded:

> It is the view of the Irish Government that the information contained in the Consultation Papers and the absence of critical information relating to primary economic factors including critical data relating to other cost factors such as transportation and security, makes it impossible for the reader to assess the justification of the [proposed MOX Plant] . . . .
>
> The Irish Government in its submissions in regard to the previous Consultation Rounds sought the unedited and full copy of the then PA Consulting Report. In the absence of this information . . . the Irish Government is reserving its right to pursue legal measures for the release of the information.[38]

35. Further, in the spring of 2001, BNFL prepared a new confidential document for departmental and ministerial consideration setting out the economic justification for the MOX Plant. Following a new public tender in April 2001, the consulting firm Arthur D. Little ("ADL") was appointed "to analyse the business case and to report on the responses to the public consultation exercise on it."[39] The terms of reference for ADL also included an instruction to form its own view as to what material should be redacted on the grounds of commercial sensitivity. ADL submitted a full version of its Report to Ministers, along with a proposed redacted public version, to which BNFL objected. The final decision about redactions in the published version was made at the Ministerial level, and the redacted ADL Report was released to the public in July 2001.[40] The transmittal letter from the Department of the Environment, Food and Rural Affairs ("DEFRA") (which had taken over responsibilities in this area from DETR) and the Department of Health stated that "[t]he published version excludes only that information whose publication would cause unreasonable damage to BNFL's commercial operations or to the economic case for the MOX plant."[41]

---

37. "UK Department of Health and DETR, British Nuclear Fuels plc. – Sellafield Mixed Oxide Plant: A Consultation Paper", *in* SMP Bundle, at Tab 4, p. 5, para. 10 (2001).

38. Ireland's Memorial, Annex 4, at No. 13.

39. Letter from UK Department of Environment, Food, and Rural Affairs ("DEFRA") and Department of Health (July 27, 2001), *in* SMP Bundle, at Tab 5.

40. United Kingdom's Counter-Memorial, para. 2.22.

41. *Ibid.*

36. A fifth public consultation ensued in August 2001. In a letter dated 7 August 2001, Ireland requested an unredacted version of the ADL Report in order "to make an independent analysis of the economic justification of the proposed [MOX] plant."[42] Ireland also restated its opposition to the MOX Plant, but did not comment in detail on the published version of the ADL Report.

37. On 3 October 2001, a decision was issued approving the manufacture of MOX at Sellafield.[43] Greenpeace, a non-governmental organization, challenged the decision in the United Kingdom courts, but its application for review was rejected,[44] and failed on appeal.[45] Ireland separately applied to the International Tribunal for the Law of the Sea ("ITLOS") for provisional measures restraining the United Kingdom from commissioning the Plant in a request which, after a hearing, was rejected.[46]

38. Against the background of these events, Ireland contended that the United Kingdom was obliged to make the information redacted from the consultation Reports available under Article 9 of the OSPAR Convention. On 15 June 2001 Ireland requested that an arbitral tribunal be constituted under Article 32 to determine its dispute with the United Kingdom concerning the United Kingdom's refusal to make available information redacted from the published versions of the PA Report and relating to the proposed MOX Plant. In its request, Ireland stated that it had previously notified the United Kingdom that a dispute had arisen as to the interpretation and application of the OSPAR Convention and that Ireland had sought to settle the dispute through bilateral diplomatic means and by raising the matter with the OSPAR Commission. A Statement of Claim was also filed.

39. In its letter dated 7 August 2001 (submitted in the context of the fifth consultation), Ireland had stated: "In the event that a copy of the full [ADL] report is not provided Ireland reserves its right to amend and extend its application in the OSPAR arbitration filed on 15 June last to include the information omitted

---

42. Ireland's Memorial, Annex 4, at No. 15.

43. Ireland's Memorial, Annex 5.

44. *See supra* note 12.

45. *Friends of the Earth Ltd. & ANR, The Queen on the Application of v. Secretary of State for the Environment, Food and Rural Affairs & ORS*, [2001] EWCA Civ. 1847.

46. *The MOX Plant Case (Ireland v. United Kingdom)*, Request for Provisional Measures, Order Dated December 3, 2001, International Tribunal for the Law of the Sea, Case No. 10. Available from www.itlos.org.

from the ADL Report."[47] By letter dated 5 September 2001, DEFRA asserted that the information excised from the public version of the ADL Report did not fall within the scope of Article 9(2).[48] By reply of 26 September 2001, the Agent for Ireland objected to DEFRA's assertion that the ADL Report did not fall within the scope of Article 9(2), and noted its intention to amend the relief sought in the Statement of Claim to include disclosure of the full unredacted ADL Report.[49]

IV. THE CLAIMS AND SUBMISSIONS OF THE PARTIES AND QUESTIONS RAISED FOR DETERMINATION BY THE TRIBUNAL

40. The formal claims of Ireland and the United Kingdom ("the Parties") were set forth in their written pleadings.

41. On the basis of Article 9 of the OSPAR Convention, Ireland, in its Memorial requested

> full disclosure of two reports commissioned by the United Kingdom Government in the context of the authorisation of a new facility and Sellafield for the production of mixed oxide (MOX) fuel . . . in order to be in a better position to consider the impacts which the commissioning of the MOX plant will or might have on the marine environment . . . [and] to be able to assess the extent of the compliance by the United Kingdom with its obligations under . . . the OSPAR Convention, the 1982 United Nations Convention on the Law of the Sea . . . and various provisions of European Community law, including in particular Council Directive 96/29 Euratom . . . .[50]

42. In its final prayer, Ireland requested the Tribunal to order and declare:

> (1) That the United Kingdom has breached its obligations under Article 9 of the OSPAR Convention by refusing to make available information deleted from the PA Report and ADL Report as requested by Ireland.
>
> (2) That, as a consequence of the aforesaid breach of the OSPAR Convention, the United Kingdom shall provide Ireland with a

47. Ireland's Memorial, Annex 4, at No. 15.

48. *Id.*, at No. 17.

49. *Id.*, at No. 16.

50. Ireland's Memorial, para. 2.

complete copy of both the PA Report and the ADL Report, alternatively a copy of the PA Report and the ADL Report which includes all such information the release of which the arbitration tribunal decides will not affect commercial confidentiality within the meaning of Article 9(3)(d) of the OSPAR Convention.

(3) That the United Kingdom pay Ireland's costs of the proceedings.

43. The United Kingdom refused to disclose the full Reports, contending in its Counter-Memorial that:

> First, Article 9 of the OSPAR Convention does not establish a direct right to receive information. Rather it requires Contracting Parties to establish a domestic framework for the disclosure of information. This the United Kingdom has done . . . .
>
> Second, in the event that the United Kingdom is wrong in this reading of Article 9, Ireland must show that the information it requests is information within the scope of Article 9(2) of the Convention. It has failed to show that this is the case . . . the information in question is insufficiently proximate to the state of the maritime area or to measures or activities affecting or likely to affect it. It is not information within the scope of Article 9(2) of the Convention . . . .
>
> Third, in the event that the United Kingdom is wrong on this point, Article 9(3)(d) of the Convention affirms the right of the Contracting Parties, in accordance with their national legal systems and applicable international regulations, to provide for a request for information to be refused on grounds of commercial confidentiality. The United Kingdom has legislated to this effect. Its refusal to disclose the particular information requested by Ireland is consistent with both national law and applicable international regulations.[51]

44. In its final prayer, the United Kingdom requested the Tribunal:

(1) to adjudge and declare that it lacks jurisdiction over the claims brought against the United Kingdom by Ireland and/or that those are inadmissible;

(2) to dismiss the claims brought against the United Kingdom by Ireland;

(3) to reject Ireland's request that the United Kingdom pay Ireland's costs, and instead to order Ireland to pay the United Kingdom's costs.

51. United Kingdom's Counter-Memorial, paras. 1.4, 1.5, 1.6.

45. It thus appears to the Tribunal that three sequential questions are raised for determination by the Tribunal, namely:

(1) Does Article 9(1) of the Convention require a Contracting Party to disclose, or to set up a procedure to disclose, "information" within the meaning of Article 9(2)?

(2) If so, does the material the disclosure of which Ireland has requested constitute "information" for the purposes of Article 9 of the Convention?

(3) If so, has the United Kingdom redacted and withheld any and what information requested by Ireland contrary to Article 9(3)(d)?

46. After a review of the procedural history of the case and the question of applicable law, the Tribunal will return to consider these questions.

V. PROCEDURAL HISTORY

47. Ireland designated Mr. David J. O'Hagan, Chief State Solicitor, as its Agent. The United Kingdom so designated Mr. Michael Wood, Legal Adviser, Foreign and Commonwealth Office ("FCO").

48. Pursuant to Article 32(4) of the OSPAR Convention, Ireland appointed Dr. Gavan Griffith QC and the United Kingdom appointed Lord Mustill as arbitrators. On 30 October 2001, they designated Professor W. Michael Reisman as chairman.

49. The Parties decided jointly to appoint the International Bureau of the Permanent Court of Arbitration ("PCA") as the registry for the arbitration and to designate The Hague as the seat of the arbitration. They also appointed Ms. Bette E. Shifman, Deputy Secretary-General of the PCA, as Registrar, and Ms. Anne Joyce, Deputy General Counsel, as Secretary, to the Tribunal.

50. On 30 November 2001, Ireland applied to the Tribunal to amend the relief it had requested in paragraph 50 of its Statement of Claim to embrace information the existence of which it was not previously aware. By letter of 7 December 2001, the United Kingdom stated that it would have no objection in principle to Ireland so extending its claim without prejudice to any question of jurisdiction or admissibility which the United Kingdom might wish to raise in due course. However, the United Kingdom contended that the manner in which Ireland had applied to amend the relief sought was "inadequate."

51. A preliminary meeting of the Tribunal with the Parties was held in London on 8 December 2001. The Parties were represented by Ms. Christina Loughlin, Deputy Agent for Ireland, and Mr. Michael Wood, and by counsel, with other Party representatives, the Registrar, and the Secretary also present.

52. At the meeting, the Party-appointed arbitrators first declared that each shared and had shared chambers with counsel appearing for both Parties, but that each believed that this fact did not constitute grounds for recusal. The Parties agreed, and had no objection.

53. The Tribunal then considered issues raised by a draft of the Rules of Procedure ("the Rules"), proposed by the Parties in accordance with Article 32(2) of the OSPAR Convention, as an alternative to the procedures outlined in Article 32(4) to (10).

54. The Tribunal also set a timetable for submissions and hearings, with Ireland's Memorial to be submitted on 8 March, the United Kingdom's Counter-Memorial on 7 June, Ireland's Reply on 19 July, and the United Kingdom's Rejoinder on 30 August 2002. It was conditionally agreed that hearings would be at The Hague during the week of 21 October 2002.

55. The Tribunal also gave leave to Ireland to amend its Statement of Claim by 15 December 2001. These amended claims were filed and served in proper form over the period late December 2001 to early January 2002.

56. Acting under Article 32(6)(b), on 12 December 2001, the Tribunal transmitted proposed Rules to the Parties for comments. Ireland had no comments. The United Kingdom proposed several changes, which were accepted by Ireland. The final text of the Rules was adopted by the Tribunal in its *Decision No. 1* of 21 February 2002.

57. The Parties filed their respective written pleadings (with statements and other documents as annexes) within the agreed time limits noted in paragraph 54 above.

58. In addition to documentary annexes, Ireland filed two successive reports by Mr. Gordon MacKerron. The United Kingdom filed successive reports by Dr. Geoffrey Varley and witness statements by Jeremy Rycroft, and one report by Mr. David Wadsworth.

59. By a joint letter of 4 October 2002, the Parties addressed several agreed procedural matters, including requirements for confidentiality and a decision that

(apart from confidential matters) the hearings should be open to the public and transcripts publicly available.

60. The Parties disagreed on the sequencing of argument and examination of witnesses. Ireland requested that each side should in turn present its entire case (including witnesses). The United Kingdom requested opening arguments to be followed by the examination of all witnesses. In separate letters dated 7 October 2002, Ireland argued that its approach reflected the traditional way in which inter-State proceedings are conducted and would assist the Tribunal in forming a view of the issues as a whole, and the United Kingdom argued that bifurcation of the hearing would facilitate the United Kingdom's cross-examination of Ireland's witness and would allow for the joinder of arguments at an earlier stage of the proceeding.

61. As a separate issue, in its letter dated 4 October 2002, Ireland requested the Tribunal to make arrangements for access by its independent counsel to the information redacted from the PA and ADL Reports. The United Kingdom replied on 8 October 2002 that this matter would best be addressed at the hearing.

62. In its letter of 8 October 2002, the United Kingdom also indicated that it might wish to refer to certain information contained in a Memorial submitted by Ireland in connection with parallel proceedings against the United Kingdom under Annex VII to the United Nations Convention on the Law of the Sea ("UNCLOS Annex VII Tribunal").

63. On 9 October 2002, these several issues were taken up in a telephone conference with the Chairman and the Agents, and counsel. The Chairman noted with approval the agreed issues advised in the Parties' joint letter of 4 October 2002. Further, it was agreed that the Parties would jointly request the UNCLOS Annex VII Tribunal to permit the disclosure in the OSPAR proceedings of material from Ireland's Memorial. This consent was later forthcoming.

64. The issue of access by counsel to the information redacted from the PA and ADL Reports was also discussed. Ireland requested that the issue be resolved before the hearing to allow Ireland sufficient time to prepare its case. The United Kingdom indicated that it was not necessarily opposed to such a review, but contended that the procedure envisioned in Article 14(4) of the Rules required one of the Parties first to tender the material. The United Kingdom also noted that there were a variety of ways of limiting disclosure of the information, including confining it to the Tribunal.

65. On 12 October 2002, the Tribunal issued its *Decision No. 2 – Conduct of the Hearings and Access to Unredacted Versions of the PA and ADL Reports*, which

directed that the hearings would be conducted over the course of eight sessions of approximately 3 hours each, between 21 and 25 October. Ireland would present its entire case (including witnesses) in the first three sessions, the United Kingdom over the following three. Each side would have one session for closing submissions.

66. Subject to several conditions, *Decision No. 2* also provided for access to the unredacted PA and ADL Reports by the Tribunal, and by independent counsel for Ireland upon them signing confidentiality undertakings. Further conditions were that counsel's access would be solely at the PCA, with no copies to be made, and that Ireland could apply to the Tribunal for permission to make the redacted material available to its other counsel and witnesses.

67. By letter dated 14 October 2002, Ireland applied for permission to make the unredacted material available to Mr. Rory Brady, the Attorney General of Ireland (who appeared as counsel for Ireland), and Mr. Gordon MacKerron.

68. By letter dated 17 October 2002, the United Kingdom proposed certain additional security arrangements for examination of the unredacted PA and ADL Reports by independent counsel for Ireland, and requested the Agent for Ireland to confirm the United Kingdom's understanding that the confidentiality undertakings provided by such independent counsel "are given also to the United Kingdom and to BNFL, and are given with the agreement of Ireland."

69. By letter dated 18 October 2002, Ireland confirmed that the confidentiality undertakings were given with the agreement of Ireland and could be taken as undertakings to the United Kingdom as well as to the Tribunal. Ireland noted that BNFL was "not a Party to these proceedings and the documents in question are the property of the United Kingdom Government and not of BNFL." With respect to the security arrangements proposed by the United Kingdom, Ireland objected to the United Kingdom's request that Ireland's independent counsel not be permitted to make any copy or note of the excised data.

70. On 18 October 2002, the eve of the hearing, the Tribunal issued its *Decision No. 3 – Procedures for Access to Unredacted Versions of the PA and ADL Reports*, which provides:

> 1. The Arbitral Tribunal acknowledges receipt of the documents tendered by the United Kingdom pursuant to paragraph 4 of its Decision No. 2 of 12 October 2002, and takes note of the letters from the United Kingdom and Ireland of 17 October 2002 and 18 October 2002, respectively.

2. In its letter, the United Kingdom requested confirmation from Ireland that it agrees to the confidentiality undertakings by independent counsel, and that such undertakings extend to the United Kingdom and to BNFL. The Arbitral Tribunal notes Ireland's statement in its letter that it agrees to the undertakings and that such undertakings extend to the United Kingdom as well as to the Tribunal. The Arbitral Tribunal further notes Ireland's agreement to abide by a Tribunal decision of confidentiality with respect to BNFL. The Tribunal notes that the undertakings of confidentiality submitted by independent counsel are general and hence avail all those whose confidential information is made available thereunder.

3. Independent counsel for Ireland may take notes during the course of its review of the above documents and may retain those notes for the duration of the hearing. However, no copies of such notes will be made, and the notes and their substance will be subject to the same confidentiality restrictions as those imposed on the unredacted copies of the PA and ADL reports. Any notes taken will be returned to the Secretary of the Tribunal at the end of the hearing and will be destroyed.

71. Prior to the opening of the hearings on Monday 21 October 2002, Ireland modified its original request concerning access to the unredacted PA and ADL Reports to exclude the Attorney General. The Agents for Ireland and the United Kingdom briefly presented arguments in closed session of the Tribunal for and against extending such access to Mr. MacKerron.

72. The Tribunal on 21 October 2002 issued its *Decision No. 4 – Access for Mr. MacKerron to Unredacted Versions of the PA and ADL Reports*, which provides:

> With respect to Ireland's application under Decision No. 2, the Tribunal does not at this stage order access to the unredacted versions of the PA and ADL reports to be given to Mr. MacKerron. If it should develop in the course of the proceeding that this presents difficulties for the Tribunal to reach a decision, it will reconsider this matter on its own initiative.

73. The hearings in the Small Court Room at the Peace Palace in The Hague ran over extended sessions from the morning of 21 to the afternoon of 25 October 2002. The Tribunal was addressed by the Agents for the Parties and by counsel as follows:

For Ireland:

Mr. Rory Brady (Attorney General)
Mr. Eoghan Fitzsimons (Senior Counsel)
Professor Philippe Sands (Counsel).

For the United Kingdom:

Dr. Richard Plender, QC (Counsel)
Mr. Daniel Bethlehem (Counsel)
Mr. Samuel Wordsworth (Counsel).

74. The following witnesses were also called and questioned by the Parties:

For Ireland:

Mr. Gordon MacKerron.

For the United Kingdom:

Dr. Geoffrey Varley
Mr. Jeremy Rycroft
Mr. David Wadsworth.

75. On 24 October 2002, the Tribunal met with Agents and counsel in chambers to consider procedural matters. After a further bench conference on that day, it was agreed that the final submissions on 24 and 25 October would be focused on the issues concerning Article 9(2) and 9(1) of the OSPAR Convention. Further, it was agreed that, if the Tribunal were to uphold any of the United Kingdom's objections, its award would be final, and no further hearings would be necessary. Alternatively, if the Tribunal were to issue an award rejecting the United Kingdom objections, a further hearing on the issues of confidentiality of the redacted information would be held (without further written submissions being filed).

76. On 24 January 2003, the Tribunal issued its *Decision No. 5 – Tribunal Request for Party Views* in which it requested the views of the Parties on an aspect of Article 9(2) that had not been addressed in the submissions:

> The third category of Article 9(2) of the OSPAR Convention refers to 'any available information . . . on activities or measures introduced in accordance with the Convention'. In their submissions, the Parties did not examine the potential relevance, if any, of this category of Article 9(2) to the question of the scope of Article 9. Without prejudice to its final

> decision on this matter, the Tribunal would appreciate the views of the Parties on this question . . . .

The Parties submitted their responses to this question within the time limits set by the Tribunal.

77. With the concurrence of the Parties, transcripts of the oral hearings (with the exception of a brief closed session wherein parts of the redacted information were discussed) as well as the filed pleadings, the Rules of Procedure, and five procedural Decisions, have been made available at the PCA website: www.pca-cpa.org.

VI. THE TRIBUNAL'S FINDINGS

78. For the reasons set out below, the Tribunal:

(i) by unanimous decision rejects the United Kingdom's request that the Tribunal find that it lacks jurisdiction over the dispute;

(ii) by unanimous decision rejects the United Kingdom's request that Ireland's claims are inadmissible;

(iii) by majority decision rejects the United Kingdom's submission that the implementation of Article 9(1) is assigned exclusively to the competent authorities in the United Kingdom and not to a tribunal established under the OSPAR Convention;

(iv) by majority decision finds that Ireland's claim for information does not fall within Article 9(2) of the OSPAR Convention; and

(v) by majority decision finds that as a consequence, Ireland's claim – that the United Kingdom has breached its obligations under Article 9 of the OSPAR Convention, by refusing, on the basis of its understanding of the requirements of Article 9(3)(d), to make available information – does not arise.

VII. APPLICABLE LAW

79. This part of the Tribunal's decision is supported by a majority comprising Professor Reisman and Lord Mustill.

1. INTERPRETATION

80. The OSPAR Convention has two authentic languages and the United Kingdom made reference to the French text of the OSPAR Convention for clarification of certain provisions. However, neither Party has alleged a discrepancy between the English and French texts for the current dispute.

81. The Parties agree that the OSPAR Convention governs the arbitration. Although the United Kingdom is Party to the Vienna Convention on the Law of Treaties ("Vienna Convention"),[52] Ireland is not, but, nonetheless, has relied upon its interpretation provisions.[53] The Parties also are agreed that the interpretation provisions of the Vienna Convention govern the construction of the OSPAR Convention.[54]

82. Articles 31 and 32 of the Vienna Convention are relevant:

> <u>Article 31 – General rule of interpretation</u>
>
> 1. A treaty shall be interpreted in good faith in accordance with the ordinary meaning to be given to the terms of the treaty in their context and in the light of its object and purpose.
>
> 2. The context for the purpose of the interpretation of a treaty shall comprise, in addition to the text, including its preamble and annexes:
>
>   (a) any agreement relating to the treaty which was made between all the parties in connection with the conclusion of the treaty;
>
>   (b) any instrument which was made by one or more parties in connection with the conclusion of the treaty and accepted by the other parties as an instrument related to the treaty.
>
> 3. There shall be taken into account, together with the context:
>
>   (a) any subsequent agreement between the parties regarding the interpretation of the treaty or the application of its provisions;

---

52. Vienna Convention on the Law of Treaties, 23 May 1969, 1155 UNTS 331.

53. Counsel for Ireland, Transcript, Day 1 Proceedings, pp. 23-24.

54. *Id.* and Counsel for the United Kingdom, Transcript, Day 2 Proceedings, p. 75.

(b) any subsequent practice in the application of the treaty which establishes the agreement of the parties regarding its interpretation;

(c) any relevant rules of international law applicable in the relations between the parties.

4. A special meaning shall be given to a term if it is established that the parties so intended.

Article 32 – Supplementary means of interpretation

1. Recourse may be had to supplementary means of interpretation, including the preparatory work of the treaty and the circumstances of its conclusion, in order to confirm the meaning resulting from the application of article 31, or to determine the meaning when the interpretation according to article 31:

(a) leaves the meaning ambiguous or obscure; or

(b) leads to a result which is manifestly absurd or unreasonable.

83. As set out in paragraph 11 above, Article 32(6)(a) of the OSPAR Convention provides that "[t]he arbitral tribunal shall decide according to the rules of international law and, in particular, those of the OSPAR Convention."[55] In dealing with general obligations, Article 2 of the OSPAR Convention provides in section 2(1)(a) that all possible steps shall be taken, "in accordance with the provisions of the Convention."

84. It should go without saying that the first duty of the Tribunal is to apply the OSPAR Convention. An international tribunal, such as this Tribunal, will also apply customary international law and general principles unless and to the extent that the Parties have created a *lex specialis*. Even then, it must defer to a relevant *jus cogens* with which the Parties' *lex specialis* may be inconsistent.

85. Ireland's submission is of a different order, namely the applicability of other conventional international law. The absence of an additional phrase in Article 2 of the OSPAR Convention (set out in para. 3 above) on the order of "and in accordance with international law" does not mean that the OSPAR Convention intended to discharge the Parties to it *inter se* from other obligations that they may have assumed under other international instruments or under general international law. However, it does mean that the competence of a tribunal established

55. Identical language is repeated in Article 19 of the Rules of Procedure for the OSPAR Tribunal.

under the OSPAR Convention was not intended to extend to obligations the Parties might have under other instruments (unless, of course, parts of the OSPAR Convention included a direct *renvoi* to such other instruments). Interpreting Article 32(6)(a) otherwise would transform it into an unqualified and comprehensive jurisdictional regime, in which there would be no limit *ratione materiae* to the jurisdiction of a tribunal established under the OSPAR Convention. Here, there is no indication that the Parties to the OSPAR Convention have, in their individual capacities, submitted themselves to such a comprehensive jurisdictional regime with respect to any other international tribunal. Nor is it reasonable to suppose that they would have accepted such a jurisdictional regime through the vehicle of the OSPAR Convention.

86. The Tribunal's interpretation is reinforced by the explicit reference in Article 9(3) of the OSPAR Convention to "applicable international regulations". The explicit incorporation of other regulations in Article 9(3) imports that, when this was not done for other provisions of the OSPAR Convention, there was no implied intention to extend the competence of the Tribunal to other parts of international conventional law.

2. THE SINTRA MINISTERIAL STATEMENT

87. Beyond the OSPAR Convention, Ireland relied in support of its claims upon the Sintra Ministerial Statement of 1998,[56] where the Ministers agreed:

> to prevent pollution of the maritime area from ionizing radiation through progressive and substantial reductions of discharges, emissions and losses of radioactive substances, with the ultimate aim of concentrations in the environment near background values for naturally occurring radioactive substances and close to zero for artificial radioactive substances.

The Ministers also noted "the concerns expressed by a number of Contracting Parties about the recent increases in technetium discharges from Sellafield and their view that these discharges should cease." The Statement continued "that the United Kingdom Ministers have indicated that such concerns will be addressed in their forthcoming decisions concerning the discharge authorisations for Sellafield." The Statement welcomed

> The announcement of the UK Government that no new commercial contracts will be accepted for reprocessing spent fuel at Dounreay, with the result of future reductions in radioactive discharges in the maritime area.

56. *See* Ireland's Memorial, Annex 8. Also available from www.ospar.org.

88. Subsequently, the OSPAR Commission (with the United Kingdom abstaining) issued its Decisions 2000/1 and 2001/1 with respect to non-reprocessing of spent nuclear fuel. By operation of Article 13, the United Kingdom is not bound by the two Commission decisions. They cannot be considered as governing law for this arbitration.

89. However, for other reasons the Sintra Statement may have created binding obligations for the United Kingdom. The International Court of Justice ("ICJ") held in the *Nuclear Tests* case that unilateral declarations accompanied by an intention to be bound may create binding obligations:

> It is well recognized that declarations made by way of unilateral acts, concerning legal or factual situations, may have the effect of creating legal obligations. Declarations of this kind may be, and often are, very specific. When it is the intention of the State making the declaration that it should become bound according to its terms, that intention confers on the declaration the character of a legal undertaking, the State being thenceforth legally required to follow a course of conduct consistent with the declaration. An undertaking of this kind, if given publicly, and with an intent to be bound, even though not made with the context of international negotiations, is binding.[57]

90. It is arguable that the United Kingdom's commitment with respect to reprocessing spent fuel at Dounreay announced in the Sintra Ministerial Statement may have created an international obligation on its part and in relation to the other states represented at the Ministerial meeting. But the question of whether the United Kingdom is under an obligation with respect to reprocessing spent fuel at Dounreay as a consequence of the announcement referred to in the Sintra Ministerial Statement is not relevant to the different question here of access to information about the activities at Sellafield.

91. The more general goals of the Sintra Ministerial Statement were plainly exhortatory. That matter aside, it appears to the Tribunal that the Sintra Ministerial Statement is not a decision or even a recommendation within the meaning of Article 13 of the OSPAR Convention.

92. Neither Article 9(1) nor Article 9(2) refers specifically to any other bodies of substantive conventional international law to which the Tribunal should have recourse.

---

57. *Case Concerning Nuclear Tests (Australia v. France)*, Judgment of 20 December 1974, 1974 ICJ Rep. 253, at p. 266, para. 43.

3. “Applicable International Regulations”

93. Article 9(3)(d) states that Parties have the right to refuse a request for information that qualifies under Article 9(2) “in accordance with their national legal systems and applicable international regulations” where the information affects “commercial and industrial confidentiality, including intellectual property . . . .”

94. Ireland acknowledged that the relevant national legal system applicable to Article 9(3) was English law.

95. Further, the Parties agreed that the 1992 Regulations (*see* para. 26 above), which give effect to Directive 90/313 (*see* para. 30 above), apply as the legislative component of the relevant national legal system. However, the Parties did not agree on their interpretation.

96. The Parties disagreed, in a number of ways, as to the reference of “applicable international regulations” in Article 9(3). Ireland contended that “applicable international regulations” means “international law and practice”.[58] The United Kingdom proposed a strict textual interpretation and submitted that there are no “applicable international regulations” for Article 9(3)(d) of the OSPAR Convention other than Directive 90/313, which was implemented in UK law.

97. On its broader submission, Ireland relied upon the Rio Declaration,[59] in particular Principle 10, and the 1998 Aarhus Convention on Access to Information, Public Participation in Decision-making and Access to Justice in Environmental Matters (the “Aarhus Convention”), which entered into force on 30 October 2001.[60] The United Kingdom replied that the Rio Declaration was not a treaty and that the Aarhus Convention has been ratified by neither Ireland nor the United Kingdom.

98. In its Reply, Ireland submitted that “‘regulations’ include all the instruments relating to the environment and access to information referred to in detail in Ireland’s Memorial . . .”, and that such instruments are to be interpreted in light of “the evolving international law and practice on access to environmental information.”[61]

---

58. Ireland’s Memorial, para. 117.

59. Declaration of the UN Conference on Environment and Development, 31 ILM 874 (1992).

60. Convention on Access to Information, Public Participation in Decision-making and Access to Justice in Environmental Matters, 25 June 1998, 38 ILM 517 (1999).

61. Ireland’s Reply, para. 42.

99. A jurisdictional clause may incorporate international law *in statu nascendi*. For example, the Special Agreement between Libya and Tunisia of 10 June 1977, submitting to the ICJ their continental shelf boundary dispute, incorporated as applicable law international maritime norms that had not yet become *lex lata*. Article 1 provided that:

> the Court shall take its decision according to equitable principles, and the relevant circumstances which characterize the area, as well as the new accepted trends in the Third Conference on the Law of the Sea.[62]

When the Parties have so empowered an international arbitral tribunal, it may apply norms that are not *lex lata*, if, in the tribunal's judgment, the norms have been accepted and are soon likely to become part of the international *corpus juris*. But the arbitral tribunal then applies them because of the Parties' instructions, not because they are "almost" law.

100. As long as it is not inconsistent with *jus cogens*, Parties may also instruct a tribunal to apply a *lex specialis* that is not part of general international law at the time. But the OSPAR Convention does not incorporate such a reference. Without such an authorization, a tribunal established under the OSPAR Convention cannot go beyond existing law. This is not to say that a tribunal cannot apply customary international law of a recent vintage, but that it must in fact be customary international law.

101. Although the issue does not arise, the Tribunal agrees with Ireland's proposal to "draw on current international law and practice in considering whether a 'commercial confidentiality' exception to a request for information may be invoked," but only insofar as such law and practice are relevant and hence admissible under Article 31(3)(c) and (d) of the Vienna Convention. However, the Tribunal has not been authorized to apply "evolving international law and practice" and cannot do so. In this regard, the Tribunal would note that the ICJ in its decision in the *Gabcíkovo-Nagymaros* case, was not, as Ireland argued,[63] proposing that it – and arguably other international tribunals – had an inherent authority to apply law *in statu nascendi*. The ICJ said:

> new norms and standards have been developed, set forth in a great number of instruments during the last two decades. Such new norms have to be taken into consideration, and such new standards given proper weight, not

62. *Case Concerning the Continental Shelf (Tunisia/Libyan Arab Jamahiriya)*, Judgment of 24 February 1982, 1982 ICJ Rep. 18, at p. 23.

63. *See* Ireland's Reply, para. 42.

> only when States contemplate new activities but also when continuing with activities begun in the past.[64]

102. The issue here is one of interpretation in good faith, as required by Article 31(1) of the Vienna Convention, if not by an essential ingredient of law itself. A treaty is a solemn undertaking and States Parties are entitled to have applied to them and to their peoples that to which they have agreed and not things to which they have not agreed.

103. Lest it produce anachronistic results that are inconsistent with current international law, a tribunal must certainly engage in *actualisation* or contemporization when construing an international instrument that was concluded in an earlier period.[65] Oppenheim, after restating the so-called law of inter-temporality (i.e., that an instrument is to be interpreted in the light of the general rules of international law in force at the time of its conclusion), adds the qualification that "in some respects the interpretation of a treaty's provisions cannot be divorced from developments in the law subsequent to its adoption."[66] But the reference in the Court's *dictum* and the doctrinal statement in Oppenheim based upon it is to developments in *law*. Wholly apart from the question of the need for actualization of a treaty made scarcely ten years earlier, the Court's reference in *Gabcíkovo-Nagymaros* is to new *law* "in a great number of *instruments*" [italics supplied] and not material that has not yet become law. As stated, a tribunal must also adjust application of a treaty insofar as one of its provisions proves inconsistent with a *jus cogens* that subsequently emerged. The present case does not raise questions of *jus cogens*.

104. For these reasons, the Tribunal cannot accept Ireland's proposal that the Aarhus Convention or that "draft proposals for a new EC Directive" be applied.[67]

105. Nonetheless, the Tribunal may apply, where appropriate, other extant international agreements insofar as they are admissible for purposes of interpretation under Article 31 of the Vienna Convention.

---

64. *Case Concerning the Gabcíkovo-Nagymaros Project (Hungary v. Slovakia)*, Judgment of 25 September 1997, 1997 ICJ Rep., at p. 7, para. 140.

65. *Legal Consequences for States of the Continued Presence of South Africa in Namibia (South West Africa) notwithstanding Security Council Resolution 276 (1970)*, Advisory Opinion, 1971 ICJ Rep., at p. 3.

66. OPPENHEIM'S INTERNATIONAL LAW, NINTH EDITION, at 1281-1282 (Sir Robert Jennings and Sir Arthur Watts eds., Longman, 1996).

67. *See* Ireland's Reply, para. 42.

VIII. FINDINGS WITH RESPECT TO ARTICLE 9(1)

1. THE PARTIES' ARGUMENTS

106. It is to be recalled that Article 9(1) provides:

> The Contracting Parties shall ensure that their competent authorities are required to make available the information described in paragraph 2 of this Article to any natural or legal person, in response to any reasonable request, without that person's having to prove an interest, without unreasonable charges, as soon as possible and at the latest within two months.

107. As noted in Part IV above, in its Counter-Memorial, the United Kingdom challenged the jurisdiction of the Tribunal and the admissibility of Ireland's claims:

> Article 9 of the OSPAR Convention does not establish a direct right to receive information. Rather it requires Contracting Parties to establish a domestic framework for the disclosure of information. This the United Kingdom has done [68]

The United Kingdom contended that Article 9(1) of the OSPAR Convention only requires Contracting Parties to "ensure that their competent authorities are required to make available the information described in Article 9(2)," and that the provision does not create a direct obligation to supply particular information. Hence, the only cause of action for a breach of Article 9 would be a failure to provide a domestic regulatory framework dealing with the disclosure of information.[69]

108. The United Kingdom also submitted that the OSPAR Convention is primarily concerned with securing adoption by Parties of programs and measures to prevent and eliminate pollution and protect the maritime environment and that once a State has done that, in accord with the OSPAR Convention, it has discharged its treaty obligations.[70]

109. In addition to its textual analysis, the United Kingdom looked to the *travaux préparatoires* which, it contended, show that Article 9(1) derives from Article 3(1) of Directive 90/313 and that the wording in the OSPAR Convention was

---

68. United Kingdom's Counter-Memorial, para. 1.4.

69. *Id.*, para. 3.4.

70. *Id.*, paras. 3.8-3.9.

amended in order to secure conformity with that directive.[71] The United Kingdom emphasized that, under Article 249 of the EC Treaty, a "Directive" is a term of art, meaning a measure which "shall be binding as to the result to be achieved, upon each Member State to which it is addressed, but shall leave to the national authorities the choice of form and methods."[72] The United Kingdom contended that, by enacting the 1992 Regulations, it had taken the legislative or administrative measures required by the OSPAR Convention, and had thereby fulfilled all of its requirements under the OSPAR Convention. It must follow that Ireland had no cause of action.[73]

110. In its Reply, Ireland submitted that Article 9(1) obliged the United Kingdom to ensure that its competent authorities make Article 9(2) information available to Ireland and that, in this instance, the competent authority is the United Kingdom Government itself.[74] In answer to the contention that the only forum in which Ireland could bring its claim with respect to non-performance under Article 9(1) would be a domestic forum, Ireland noted that the concession by the United Kingdom that Ireland would be entitled to complain to the European Commission or the European Court of Justice ("ECJ") confirmed that such a violation of the OSPAR Convention is actionable at the international level. Moreover, Ireland contrasted Article 4 of Directive 90/313, from which Directive Article 9(1) was drawn, which specifically directs a person who considers that his request has been unreasonably refused to seek judicial or administrative review in accordance with the relevant national legal system, with Article 9(1), which had no such limitations on recourse to national decision makers.[75]

111. Ireland further contended that the obligation of Contracting Parties to "ensure that their competent authorities are required to make available" certain information, this provision constitutes an "obligation of result," rather than an obligation, as the United Kingdom submitted, "to provide for a domestic regulatory framework dealing with the disclosure of information."[76] Ireland further argued that its case

71. *Id.*, para. 3.9.

72. *Id.*, para. 3.11.

73. *Id.*, paras. 3.11-3.13.

74. Ireland's Reply, para. 7.

75. *Id.*, paras. 8-9.

76. Counsel for Ireland, Transcript, Day 1 Proceedings, p. 39.

was not "premised on the basis that it has a direct right to the information under Article 9(1)."[77]

112. Ireland also asserted that

- Ireland is a "natural or legal person" within the meaning of Article 9(1) of the OSPAR Convention,[78] and

- The entity to which Ireland's request was directed, DETR, and its successor entity, DEFRA, are "competent authorities," also within the meaning of Article 9(1).[79]

113. With respect to the possibility of recourse to domestic courts, Ireland stated:

> Article 9 of the OSPAR Convention is an unincorporated convention; it is not part of English law and it cannot be invoked before the English courts.[80]

Regarding possible alternative legal actions involving the Commission of the European Communities or the ECJ, Ireland argued that:

- While such approaches may be available to Ireland, it is up to Ireland to decide in which forum to seek a remedy,[81] and

- The United Kingdom's recognition of a potential cause of action before the ECJ under the freedom of information provisions of Directive 90/313 undercuts the United Kingdom's contention that Ireland has no cause of action before the Tribunal under the nearly identical language of Article 9 of the OSPAR Convention.[82]

114. The United Kingdom argued in answer that the requirement to make information available "to any natural or legal person"

---

77. Counsel for Ireland, Transcript, Day 1 Proceedings, p. 38.

78. *Ibid.*

79. Counsel for Ireland, Transcript, Day 1 Proceedings, pp. 38-39; Day 4 Proceedings, pp. 14-23.

80. Counsel for Ireland, Transcript, Day 1 Proceedings, p. 41.

81. Counsel for Ireland, Transcript, Day 4 Proceedings, pp. 18-24.

82. Ireland's Reply, paras. 8-9; Counsel for Ireland, Transcript, Day 4 Proceedings, pp. 21-22.

> . . . makes no sense if the obligation on the Contracting Party is to make available specific information on request. This is a treaty. A natural or legal person other than an OSPAR Party has no standing under the Convention.[83]

115. The United Kingdom also submitted that its interpretation of Article 9(1) is dictated by the need to give effect to all the words in the provision; is consistent with the language and structure of Article 9(3); and also is consistent with the object and purpose of the OSPAR Convention, namely "the adoption by Contracting Parties of programmes and measures to prevent and eliminate pollution and protect the maritime area."[84]

116. In its Rejoinder, the United Kingdom distinguished between Ireland's right to receive information under domestic law and its right under international law.[85] According to the United Kingdom, Ireland is entitled to repair to this Tribunal to secure United Kingdom compliance with its obligations under Article 9 to ensure that its competent authority shall be required to make available the kind of information found in Article 9(2). But if the United Kingdom legislation (and, presumably, its administrative implementation) is sufficient, under the OSPAR Convention, a grievance of Ireland with respect to a particular decision must be resolved at the national level.[86]

117. The United Kingdom also revisited its reference to the *travaux préparatoires* (noted in para. 109 above), noting the relationship between the drafting of Article 9(1) and (2) and the language of Directive 90/313 to argue that, as a matter of both the plain language of the Directive as well as EU law and practice, the Directive was clearly not intended to create a direct right of access to information by EU Member States. In the United Kingdom's view, it followed from these circumstances that the nearly identical formulation adopted in Article 9 signalled the Contracting Parties' intention to create a similarly circumscribed obligation, namely one limited to putting in place the required domestic legislation.[87]

---

83. Counsel for the United Kingdom, Transcript, Day 2 Proceedings, p. 76.

84. *Id.*, pp. 76-77.

85. United Kingdom's Rejoinder, para. 12.

86. *Id.*, para. 14.

87. *Id.*, para. 13.

2. THE TRIBUNAL'S DECISION WITH RESPECT TO JURISDICTION UNDER ARTICLE 9(1)

118. The United Kingdom has characterized its objection to Ireland's claim under Article 9(1) as going to the lack of jurisdiction of the Tribunal and/or being inadmissible. However, in the unanimous view of the Tribunal the question posed by Ireland with respect to Article 9(1) is not one of jurisdiction or admissibility, but one of substance, namely what is the purport of Article 9(1) under the facts of this case.

119. The remaining holding with respect to Article 9(1) is supported by a majority comprising Dr. Griffith and Lord Mustill.

120. As noted above (para. 109), Regulation 3 of the 1992 Regulations giving effect to Directive 90/313 is relied upon by the United Kingdom as constituting its compliance under domestic law with the requirements of Article 9(1).[88] The United Kingdom contends that the mandated regime under domestic law is not required to be expressed as being pursuant to the OSPAR Convention obligation.[89] The Tribunal agrees that the standard may be satisfied in a form such as the 1992 Regulations, which are otherwise justified under Directive 90/313.

121. Although, as asserted by the United Kingdom and not contested by Ireland, it would be a proper subject for this Tribunal's jurisdiction, it is no part of Ireland's claims in this dispute that there are defects within the domestic regime to the extent that the 1992 Regulations fall below the standards required by Article 9(1).

122. For the purpose of this issue of construction of Article 9(1), the Tribunal assumes that the redacted information sought by Ireland is of a sort required to be disclosed.

123. The issue remains one of interpretation of public international law, namely whether, as the United Kingdom contends, the obligation of a Contracting Party under Article 9(1) is completely discharged by putting in place an appropriate domestic regulatory framework so that disputes about specific applications of the obligations under Article 9 are to be exclusively determined within the municipal law of the Contracting Party. Should this be the case, the appropriate forum for Ireland with respect to its claims that information to which it was entitled under the OSPAR Convention was improperly withheld will be found in the United Kingdom municipal system.

---

88. *See* United Kingdom's Counter-Memorial, para. 3.12.

89. Counsel for the United Kingdom, Transcript, Day 2 Proceedings, p. 85.

124. If Article 9(1) is to be interpreted as maintained by Ireland, then this Tribunal may exercise its jurisdiction to consider the merits of the refusal of the United Kingdom's competent authorities to disclose information contained in the PA and ADL Reports, provided that such information falls within the definition of Article 9(2) of the OSPAR Convention.

125. Consistently with Article 31 of the Vienna Convention, the Parties have focused their arguments on the treaty text to determine the meaning of the Article 9(1) obligation.[90] The Tribunal applies this approach to examine the terms of Article 9(1) in the context of the entire Article 9 and the OSPAR Convention.

126. The Tribunal first examines the meaning of the obligation in the context of the OSPAR Convention regime, taking into account its objects and purposes and also the fact that a dispute settlement clause is incorporated by Article 32. In confirmation of this analysis the Tribunal also is guided by Article 32(6)(a) to analyze the relevant rules of international law that inform the meaning of the obligation of Article 9(1), and in particular (the now superseded) Directive 90/313.

127. Article 9 is an access to information provision that must be taken to articulate the Contracting Parties' intentions as expressed within the framework of the general objectives and the particular other provisions of the OSPAR Convention. As much as do the other operative articles of the OSPAR Convention, the disputes clause, Article 32, applies Article 9 as an enforceable obligation in its particular subject matter. Its provisions for disclosure of defined information must be taken to have an intended bite beyond being an expression of aspirational objectives for the domestic laws of the Contracting Parties.

128. The main purpose of the OSPAR Convention is the protection of the marine environment and the elimination of the marine pollution in the North-East Atlantic. The objectives of the OSPAR Convention are set out in its Preamble and include, *inter alia*, obligations

- to protect the marine and other environments;
- to prevent and eliminate pollution;
- to prevent and punish infringements;
- to assist a Contracting Party;

90. *See, e.g.*, United Kingdom's Counter-Memorial, paras. 3.1-3.3; United Kingdom's Rejoinder, para. 14; Counsel for Ireland, Transcript, Day 1 Proceedings, pp. 23-24.

- to conduct research; and
- to prevent dumping.

129. For the achievement of these aims the framers of the OSPAR Convention have carefully applied differential language to provide for stipulated levels of engagement of treaty obligation to achieve these objectives. There is a cascading standard of expression providing for the particular obligations imposed on a Contracting Party. For example, there are mandatory provisions that provide for Contracting Parties:

- to take some act ("shall apply", "shall include", "shall undertake", "shall co-operate" or "shall keep");
- actively to work towards an objective ("take all possible steps", "implement programs", "carry out programs");
- to deal with issues of planning for the objective ("establish programs", "adopt", "define", "draw up", "develop", "take account of"); and
- to take measures ("take", "adopt", "plan", "apply", "introduce", "prescribe", "take into account").

At a lesser level of engagement, other provisions provide for information to be dealt with ("collect", "access information") or that systems be set up ("provide for", "establish").

130. When read as a whole (including the Annexes), it is plain to the Tribunal that the entire text discloses a carefully crafted hierarchy of obligations or engagement to achieve the disparate objectives of the OSPAR Convention. Those who framed the OSPAR Convention expressed themselves in carefully chosen, rather than in loose and general, terms. They plainly identified matters for mandatory obligation for action by Contracting Parties, as in

- Article 5 ("The Contracting Party shall take");
- Article 6 ("The Contracting Party shall . . . undertake");
- Article 7 ("The Contracting Party shall co-operate . . ."); and
- Article 8 ("The Contracting Party shall establish . . .", "The Contracting Party shall have regard . . .").

131. Further, requirements for Contracting Parties to ensure a result are not confined to Article 9(1). Importantly, the general obligations expressed in Article 8(2) and embraced under Article 2(3)(b)(ii) are that the Contracting Parties shall ". . . ensure the application of best available techniques and best environmental practice . . . ." Similarly, Article 4(1) of Annex II dealing with dumping requires that the Contracting Parties shall "ensure" the required result and, under Article 10(1), shall "ensure compliance" by vessels or aircraft. Likewise, Article 5(1) of Annex III demands that the Contracting Parties shall "ensure" that their competent authorities implement the relevant applicable decisions, recommendations, and all other agreements adopted under the OSPAR Convention.

132. The issue for determination is whether the requirement in Article 9(1) "to ensure" the obligated result, mandates a result rather than merely a municipal law system directed to obtain the result.

133. In the context of the language used within Article 9, it remains for the Tribunal to discern the extent of the comprised obligation. Whatever its particular replication of Directive 90/313, what does appear plain to the Tribunal is that the obligation expressed in Article 9(1) by the requirement that a Contracting Party "shall ensure" the stipulated result is a reflection of a deliberate rather than a lax choice of vocabulary. It illustrates the application of a chosen (and strong) level of expression, deftly applied by the drafters to the particular and, to them, important subject matter of disclosure of information to any persons, whether nationals or not, who request it. It is expressed at the higher level of obligation, and when applying it in the complex of the provisions on disclosure of information embraced by the scheme of Article 9, the Tribunal sees no reason to read its particular language in a way that is discordant with the structure and use of language in the entire OSPAR Convention. The search is for conformity of meaning within the OSPAR Convention.

134. On that approach, the Tribunal finds that the obligation is to be construed as expressed at the mandatory end of the scale. The applied requirement of Article 9(1) is read by the Tribunal as imposing an obligation upon the United Kingdom, as a Contracting Party, to ensure something, namely that its competent authorities "are required to make available the information described in paragraph 2 . . . to any natural legal person, in response to any reasonable request."

135. It appears to the Tribunal that to accept the expression of the requirement "to ensure" a result as expressed at the lesser level of setting up a regime or system directed to obtain the stipulated result under the domestic law of the Contracting Party, as is contended by the United Kingdom, would be to apply an impermissible gloss that does not appear as part of the unconditional primary obligation under Article 9(1). In contrast, a limitation of this sort is expressly

embraced in the scheme of Article 9(3) providing for exceptions of disclosure expressed by reference to criteria to be imposed by the Contracting Parties "in accordance with their national legal systems." The fact that Article 9(3) engages such a limitation by reference to domestic law forecloses the possibility that Article 9(1) silently and similarly limits the obligation upon a Contracting Party to that of putting in place a domestic legal regime providing for disclosure in compliance with the Article 9 obligations.

136. A further matter that militates in favor of this interpretation is the fact that Article 9(1) identifies the objective criteria that should be met when a request to provide information is received by the competent authorities of a Contracting State. Hence, compliance by a Contracting State with these criteria may itself become a separate subject matter of arbitration under Article 32.

137. For these reasons in this aspect it appears to the Tribunal that Article 9(1) is advisedly pitched at a level that imposes an obligation of result rather than merely to provide access to a domestic regime which is directed at obtaining the required result.

138. In adopting this construction the Tribunal gives full effect to the terms of Article 9(1), including particularly the requirement that as a Contracting Party the United Kingdom "shall ensure that their competent authorities are required to make available the information." The Tribunal applies, rather than excises, this clause as the defining part of the obligation.

139. The Tribunal derives further support for its mere textual analysis of Article 9(1) from the relevant rules of international and European Union law.

140. The Parties are in agreement on the origins of Article 9(1) as derived from, and closely following, the language of Directive 90/313.[91] As noted above, in support of its position the United Kingdom refers to the notion of a directive as defined in Article 249 of the EC Treaty as a measure which "shall be binding as to the result to be achieved, upon each Member State to which it is addressed, but shall leave to the national authorities the choice of form and methods."[92] The United Kingdom submits that by adopting the language of the Directive, the Contracting Parties to the OSPAR Convention evinced their intention to adopt the same

91. United Kingdom's Counter-Memorial, para. 3.9. *See also* Counsel for Ireland, Transcript, Day 1 Proceedings, pp. 24-25; Counsel for the United Kingdom, Transcript, Day 2 Proceedings, pp. 78-79; Counsel for Ireland, Transcript, Day 4 Proceedings, p. 21.

92. *See* Article 249 of the Treaty Establishing the European Community ("EC Treaty"), 2002 OJ (C 325), as cited in the United Kingdom's Counter-Memorial, para. 3.11.

approach,[93] namely that a State's only obligation is to take such legislative or administrative measures as may be appropriate to achieve the stated objective.[94]

141. In considering these contentions the Tribunal first notes that the adoption of a similar or identical definition or term in international texts should be distinguished from the intention to bestow the same normative status upon both instruments. The complex of instruments whose wording was used by the drafters may include unilateral statements, position papers, declarations, recommendations, and the like. While the language of such sources might be instrumental to the extent that it allows one to trace and understand the origins of specific treaty terms, their normative value should not be attributed to similarly worded legal obligations imposed by that treaty. As the ITLOS has helpfully observed in its Order of 3 December 2001:

> [E]ven if the OSPAR Convention, the EC Treaty and the Euratom Treaty contain rights or obligations similar to or identical with the rights and obligations set out in [UNCLOS], the rights and obligations under those agreements have a separate existence from those under [UNCLOS].

Further,

> [T]he application of international law rules on interpretation of treaties to identical or similar provisions of different treaties may not yield the same results, having regard to, *inter alia*, differences in the respective contexts, objects and purposes, subsequent practice of parties and *travaux préparatoires*.[95]

142. Each of the OSPAR Convention and Directive 90/313 is an independent legal source that establishes a distinct legal regime and provides for different legal remedies. The United Kingdom recognizes Ireland's right as an EU Member State to challenge the implementation of the Directive in the United Kingdom's domestic legal system before the ECJ.[96] Similarly, a Contracting Party to the OSPAR Convention, with its elaborate dispute settlement mechanism, should be

93. *Ibid.*

94. United Kingdom's Rejoinder, para. 13.

95. *Id.*, para. 51.

96. United Kingdom's Counter-Memorial, paras. 3.13-3.15.

able to question the implementation of a distinct legal obligation imposed by the OSPAR Convention in the arbitral forum, namely this designated Tribunal.[97]

143. Pursuant to Article 4 of Directive 90/313, legal action against a State in breach is to be pursued domestically. However, and in contrast, the OSPAR Convention contains a particular and self-contained dispute resolution mechanism in Article 32, in accordance with which this Tribunal acts. Article 9(1) does not provide for an exception to the OSPAR disputes clause by referring, for instance, to an exclusive municipal remedy, and is therefore as subject to review by an arbitral tribunal as any other provision of the OSPAR Convention. The similar language of the two legal instruments, as well as the fact that the 1992 Regulations are an implementing instrument for both Directive 90/313 and the OSPAR Convention, does not limit a Contracting Party's choice of a legal forum to only one of the two available, i.e. either the ECJ or an OSPAR tribunal. Nor, contrary to the United Kingdom's contention, does it suggest that the only cause of action available to Ireland is confined exclusively to those provided for by Directive 90/313 and implementing legislation. The primary purpose of employing the similar language is to create uniform and consistent legal standards in the field of the protection of the marine environment, and not to create precedence of one set of legal remedies over the other.

144. The proposed reading of Article 9(1) also is consistent with contemporary principles of state responsibility. A State is internationally responsible for the acts of its organs. On conventional principles, a State covenanting with other States to put in place a domestic framework and review mechanisms remains responsible to those other States for the adequacy of this framework and the conduct of its competent authorities who, in the exercise of their executive functions, engage the domestic system.

145. Amongst others, this submission is confirmed by Articles 4 and 5 of the International Law Commission Draft Articles on the Responsibility of States for Internationally Wrongful Acts,[98] providing for rules of attribution of certain acts to States. On the international plane, acts of "competent authorities" are considered to be attributable to the State as long as such authorities fall within the

---

97. In 2001 the ITLOS was confronted with a similar situation. In response to the jurisdictional objections raised by the United Kingdom, it remarked that "since the dispute before the Annex VII arbitral tribunal concerns the interpretation or application of the [UNCLOS] and no other agreement, only the dispute settlement procedures under the Convention are relevant to that dispute." *See The Mox Plant Case*, *supra* note 46, para. 52.

98. *International Liability for injurious consequences arising out of acts not prohibited by international law (prevention of transboundary harm from hazardous activities)*, Report of the International Law Commission, 53rd Session, Supp. No. 10, UN Doc. A/56/10, 366 (2001).

notion of state organs or entities that are empowered to exercise elements of the governmental authority. As the ICJ stated in the *LaGrand* case, "the international responsibility of a State is engaged by the action of the competent organs and authorities acting in that State, whatever they may be."[99]

146. It follows as an ordinary matter of obligation between States, that even where international law assigns competence to a national system, there is no exclusion of responsibility of a State for the inadequacy of such a national system or the failure of its competent authorities to act in a way prescribed by an international obligation or implementing legislation. Adopting a contrary approach would lead to the deferral of responsibility by States and the frustration of the international legal system.

147. In support of its interpretation of Article 9(1), Ireland invoked the *LaGrand* case, to contend that the ICJ found that Article 36(1)(b) of the Vienna Convention on Consular Relations[100] created an obligation of result, and that "the failure to provide consular access at the national level gave rise to a dispute over which the International Court of Justice had jurisdiction."[101] Although there are obvious differences in the direct and indirect references to the relevant competent authorities between Article 9(1) of the OSPAR Convention and Article 36(1)(b) of the Vienna Convention on Consular Relations, one Tribunal member (Dr. Griffith) finds some independent support in *LaGrand* for these conclusions. However, the Tribunal's position on the more direct issues of textual interpretation make it unnecessary to invoke such other matters of confirmatory support.

148. For these reasons the Tribunal rejects the contention of the United Kingdom based on Article 9(1), and determines that upon its proper construction Article 9(1) requires an outcome of result, namely that information falling within the meaning of Article 9(2) (and not excluded under Article 9(3)) is in fact disclosed in conformity with the Article 9 obligation imposed upon each Contracting Party.

---

99. *LaGrand (Germany v. United States of America)*, Provisional Measures, Order of 3 March 1999, 1999 ICJ Rep. 9, at p. 16, para. 28.

100. Vienna Convention on Consular Relations, 24 April 1963, 596 UNTS 261.

101. Counsel for Ireland, Transcript, Day 1 Proceedings, p. 43. Ireland also submitted that "the ICJ made it clear that its function was to review the merits of whether the United States had complied with obligations to ensure consular access to an individual in the United States, a German national." *Id.*, at p. 35.

IX. FINDINGS WITH RESPECT TO ARTICLE 9(2)

1. THE PARTIES' ARGUMENTS

149. In its Memorial, Ireland proposed a broad scope to Article 9(2), so as not to require a Party claiming information under Article 9(1) to demonstrate that the information sought relates directly to activities that adversely affect the maritime area. Ireland contended:

> Ireland submits that this is the wrong approach. The correct approach is to look at the information *as a whole*. The purpose of the PA and ADL Reports is to examine the justification of the MOX Plant. That Plant makes possible an activity – MOX Production – which will undoubtedly have an adverse impact on the maritime area covered by the OSPAR Convention. The omitted information relates closely to various important aspects of that activity, and contributed to the determination whether that activity should be permitted. It would be unduly restrictive to read Article 9(2) as referring only to *environmental* information about such an activity.[102]

Thus Ireland claimed that "every aspect of the omitted information [in the PA and ADL Reports] is covered by Article 9(2)."[103]

150. Ireland relied, for its proposed interpretation, on the broad definition of "environmental information" in the Aarhus Convention. That treaty, Ireland argued, is not a progressive development in the law concerning access to environmental information. Rather, the reference to "cost-benefit and other economic analyses and assumptions used in environmental decision-making" in Article 2(3) of the Aarhus Convention "makes clear that which was implicit" in the OSPAR Convention,[104] and should, Ireland submitted, be applied by virtue of Article 31(3)(c) of the Vienna Convention.[105] Ireland contended that this interpretation was applied by the ECJ in *Mecklenburg v. Kreis Pinneberg – Der Landrat* ("*Mecklenburg*").[106]

151. Ireland contended further that it is clear from the context in which the Reports were prepared – notably, that they were commissioned by United Kingdom

---

102. Ireland's Memorial, para. 98.

103. *Id.*, at para. 99.

104. Counsel for Ireland, Transcript, Day 1 Proceedings, p. 52.

105. Ireland's Memorial, para. 101.

106. C-321/96, *Mecklenburg v. Kreis Pinneberg – Der Landrat*, [1999] 2 CMLR 418, 435.

Government departments of state with responsibilities for the environment, apparently to comply with Directive 90/313 and its implementing regulations – that they contain information which is appropriately labeled "environmental." Ireland argued that it

> does not dispute . . . that the information relates to a commercial activity, the operation of the MOX plant . . . . But the question of whether the information has a commercial character is not dispositive of whether it falls within Article 9(2) . . . . It is self evident, we say, that the PA and ADL reports are information on activities or measures adversely affecting or likely to affect the maritime area of the Irish Sea.[107]

152. In response to the United Kingdom's argument that the information requested is not "directly and proximately related" to activities or measures adversely affecting or likely to affect the maritime area, Ireland argued that such a test is not part of Article 9(2), and that, in any event, "without the ADL report there would be no discharges from the MOX plant into the Irish Sea. It is hard to think how that report cannot even according to that test be direct and proximate."[108]

153. In its Counter-Memorial, the United Kingdom submitted that Ireland's contention "that all information relating to the production of MOX will be information that comes within the scope of Article 9(2)" does not address the relevant question.[109] According to the United Kingdom, that relevant question is not whether MOX production will affect the maritime area, because "all such information has been in the public domain for many years."[110] According to the United Kingdom, the information which is contemplated in Article 9(2) is "information on activities or measures adversely affecting or likely to affect the maritime area."[111] This narrower reading prevents the OSPAR Convention from being used to conduct "fishing expeditions."[112] Thus, the United Kingdom argued

---

107. Counsel for Ireland, Transcript, Day 1 Proceedings, p. 48.

108. *Id.*, p. 50.

109. United Kingdom's Counter-Memorial, para. 4.2.

110. *Id.*, para. 4.7.

111. *Id.*, para. 4.3.

112. *Ibid.*

> Article 9 of the OSPAR Convention cannot be used to secure disclosure of any and all information concerning undertakings some aspects of the operations of which may touch upon the state of the maritime area.[113]

The United Kingdom contended that, for the purpose of identifying information covered by Article 9, one must identify the relevant activity "adversely affecting or likely to affect" the maritime area to which the information must relate. "In the instant case, the relevant activity would be the discharging of radioactive elements from the MOX Plant into the maritime area," and not the MOX Plant per se, "which in many of its aspects has nothing whatsoever to do with the maritime area."[114]

154. The United Kingdom asserted that the OSPAR Convention is not a "freedom of information treaty," and if it "had been intended to secure disclosure of any and all information of whatever nature on an activity whose operations affected or potentially affected the state of the maritime area, clearer language would have been used."[115]

155. With respect to the definition (cited by Ireland) of "environmental information" in Article 2(3) of the Aarhus Convention, the United Kingdom stated that this treaty is not in force for either Ireland or the United Kingdom, nor have its provisions been adopted in an EC directive. Moreover, the United Kingdom argued, the language in the Aarhus Convention concerning economic information used in environmental decision making is "new" and is perceived that way in a number of official documents published by the European Union and United Kingdom Government.

156. The United Kingdom argued that the *Mecklenburg* case is not on point, particularly because the ECJ in that case was interpreting language from a part of Directive 90/313 that does not have a precise equivalent in the OSPAR Convention.

157. In response to the question posed to the Parties by the Tribunal in *Decision No. 5*, Ireland stated that the words "activities and measures" in the third category of Article 9(2) confirm that the words in the second category of Article 9(2) are to be given a "broad meaning" and that "[t]he third category may therefore assist in the proper interpretation of the extent of the type of information envisaged by

---

113. *Id.*, para. 4.4.

114. Counsel for the United Kingdom, Transcript, Day 2 Proceedings, p. 86.

115. *Id.*, at p. 87.

Article 9(2), including the scope of the first and second categories."[116] Ireland also submitted that the words

> 'activities or measures introduced in accordance with the Convention' implies that the second category includes (but is not limited to 'activities or measures not in accordance with the Convention.') [underlining in original]

Finally, Ireland submitted that the third category supported the view that "the drafters foresaw and provided for one State Party being entitled to make a request to another State Party under Article 9."[117]

158. In response to the question posed to the Parties by the Tribunal in *Decision No. 5*, the United Kingdom submitted that the third category of Article 9(2), "like the first category, forms part of the context to be taken into account in interpreting the second category" and that "there must be a direct relationship between the information, on the one hand, and the state of the maritime area (as defined in Article 1(a) of the OSPAR Convention), on the other."[118]

2. THE TRIBUNAL'S DECISION WITH RESPECT TO THE CLAIMS RELATING TO ARTICLE 9(2)

159. The United Kingdom has characterized its objection to Ireland's claim under Article 9(2) as going to the lack of jurisdiction of the Tribunal and/or being inadmissible. In the unanimous view of the Tribunal, however, the question posed by Ireland with respect to Article 9(2) is not one of jurisdiction or admissibility, but one of substance, viz. what is the purport of Article 9(2) under the facts of this case.

160. The remaining holding with respect to Article 9(2) is supported by a majority comprising Professor Reisman and Lord Mustill.

161. The Tribunal has not been requested to issue an advisory opinion as to the abstract meaning of Article 9(2) of the OSPAR Convention, but rather to apply the provision to a specific controversy about 14 categories of information

---

116. Letter from the Agent for Ireland to the Secretary of the Tribunal (February 21, 2003), on file at the PCA.

117. *Ibid.*

118. Letter from the Agent for the United Kingdom to the Secretary of the Tribunal (February 21, 2003), on file at the PCA.

redacted from the PA and ADL Reports. In its Memorial, Ireland identified those 14 categories as information relating to:

(A) Estimated annual production capacity of the MOX facility;

(B) Time taken to reach this capacity;

(C) Sales volumes;

(D) Probability of achieving higher sales volumes;

(E) Probability of being able to win contracts for recycling fuel in 'significant quantities';

(F) Estimated sales demand;

(G) Percentage of plutonium already on site;

(H) Maximum throughput figures;

(I) Life span of the MOX facility;

(J) Number of employees;

(K) Price of MOX fuel;

(L) Whether, and to what extent, there are firm contracts to purchase MOX from Sellafield;

(M) Arrangements for transport of plutonium to, and MOX from, Sellafield;

(N) Likely number of such transports.[119]

It will be recalled that in its Amended Statement of Claim, the first relief which Ireland sought was an order and declaration that the United Kingdom had breached its obligations under Article 9 of the OSPAR Convention "by refusing to make available information deleted from the PA Report and the ADL Report." Ireland's second prayer for relief was, in effect, for an order for the provision by the United Kingdom of those parts of the PA and ADL Reports that had been redacted or, contingently, those parts that had been redacted but that did not affect commercial confidentiality within the meaning of Article 9(3)(d). The specific issue before the Tribunal is whether the redacted portions of the PA and ADL Reports, viewed as categories, constitute "information" within the meaning of Article 9(2). The Tribunal distinguishes here between the categories of redaction and the content of those categories. A determination under Article

119. Ireland's Memorial, para. 75. Ireland also provided more detailed lists of specific items deleted from the PA and ADL Reports in its Memorial Annexes 3 and 3 B, respectively.

9(3)(d) would require a detailed examination of the *content* of the various categories of redaction. A determination under Article 9(2) requires only an examination of the categories of redaction, in order to determine whether they fall within the definition of "information" in Article 9(2).

162. As will be recalled, Article 9(2) provides:

> The information referred to in paragraph 1 of this Article is any available information in written, visual, aural or data-base form on the state of the maritime area, on activities or measures adversely affecting or likely to affect it and on activities or measures introduced in accordance with the Convention.

163. Article 9(2), whose chapeau is "Access to Information," establishes the scope of information to which, subject to specific enumerated rights of refusal in Article 9(3), the obligation in Article 9(1) relates. The scope of the information in the provision is not environmental, in general, but, in keeping with the focus of the OSPAR Convention, "the state of the maritime area." It is manifest to the Tribunal that none of the above 14 categories in Ireland's list can plausibly be characterized as "information . . . on the state of the maritime area." The Tribunal could, thus, rest its decision on the fact that none of the material in the 14 categories falls within the definition of "information" in Article 9(2).

164. In response to this, Ireland's submission of what might be called an interpretative theory of "inclusive causality" would overcome this difficulty. Ireland argued, it will be recalled,

> without the ADL report there would be no discharges from the MOX plant into the Irish Sea. It is hard to think how that report cannot even according to that test be direct and proximate.[120]

Under an interpretative theory of inclusive causality, anything, no matter how remote, which facilitated the performance of an activity is to be deemed part of that activity. Legislators and drafters of treaties may adopt a theory of inclusive causality. The question is whether the drafters of the OSPAR Convention did. Some parts of Article 9(2) are, indeed, quite expansive, but other parts make abundantly clear that while the drafters sought inclusiveness with respect to some aspects of the information covered by Article 9(2), they had no intention of adopting a theory of inclusive causality. The Tribunal now turns its attention to these matters.

---

120. Counsel for Ireland, Transcript, Day 1 Proceedings, p. 50.

165. Article 9(2) identifies three categories, within each of which "any available information" falls within the obligations of Article 9, unless that category has a restriction. The drafters' selection of the adjectives "any" and "available" in Article 9(2) is significant.

166. The adjective "available" indicates that the drafters were not imposing an obligation on a Contracting Party to gather and process information of a certain sort upon the request of any natural or legal person, but rather were limiting the obligation of the Contracting Parties under Article 9 to information which had already been gathered and was already available to them. This provision is thus similar, in effect, to Article 14(1) of the Convention on Civil Liability for Damage Resulting from Activities Dangerous to the Environment of 1993,[121] which establishes access to information simply if it is "held by public authorities." In this respect, the obligation of Article 9(2) differs from obligations in certain national instruments, under which a claimant with standing may require that a government or its agency or instrumentality gather, process, and make available certain types of information.

167. The adjective "any" indicates that, unless set out explicitly within the three categories enumerated in Article 9(2), no selections or restrictions are implied. One such explicit class of restrictions is to be found in the rights of refusal to a request for information under the grounds specified in Article 9(3). Apart from exceptions, the insertion by the drafters of the adjective "any" requires an applier to interpret extensively *within* each of the three categories. Once a matter is found to fall within one of the categories of Article 9(2), the presumption is that it is within the scope of the OSPAR Convention. This mandate for an extensive construction of the provision is reinforced by the drafters' selection of the term "information."

168. Article 1 does not define "information" but it is clear that it is a broad and inclusive reference with respect to the state of the maritime area. The point of emphasis, however, is that it is "information" about the state of the maritime area. The three categories of "information . . . on the state of the maritime area" in Article 9(2) are

(i) "any available information" on "the state of the maritime area,"

(ii) "any available information" on "activities or measures adversely affecting or likely to affect . . . the maritime area,"

121. Convention on Civil Liability for Damage Resulting from Activities Dangerous to the Environment, 21 June 1993, ETS 150.

(iii) "any available information" on "activities or measures introduced in accordance with the Convention."

169. In their submissions to the Tribunal, both Parties focused attention on the second category of Article 9(2). In their responses to the Tribunal's *Decision No. 5* requesting their views on the third category, both Parties again indicated that the critical category for decision was the second. Accordingly, the Tribunal will direct its attention to this category, relying on the other categories, insofar as appropriate, for purposes of interpreting the second, as did the Parties in their responses.

170. It is clear that Article 9(2) is not a general freedom of information statute. The information here is restricted in a number of ways. First, as noted, it is restricted by the term "maritime area," which appears in the first and second categories of Article 9(2), and is given a specific definition in Article 1(a) and 1(b), which provide

> (a) 'Maritime area' means the internal waters and the territorial seas of the Contracting Parties, the sea beyond and adjacent to the territorial sea under the jurisdiction of the coastal state to the extent recognised by international law, and the high seas, including the bed of all those waters and its sub-soil, situated within the following limits:
>
> (i) those parts of the Atlantic and Arctic Oceans and their dependent seas which lie north of 36° north latitude and between 42° west longitude and 51° east longitude, but excluding:
>
> (1) the Baltic Sea and the Belts lying to the south and east of lines drawn from Hasenore Head to Gniben Point, from Korshage to Spodsbjerg and from Gilbjerg Head to Kullen,
>
> (2) the Mediterranean Sea and its dependent seas as far as the point of intersection of the parallel of 36° north latitude and the meridian of 5° 36' west longitude;
>
> (ii) that part of the Atlantic Ocean north of 59° north latitude and between 44° west longitude and 42° west longitude.
>
> (b) 'Internal waters' means the waters on the landward side of the baselines from which the breadth of the territorial sea is measured, extending the case of watercourses up to the freshwater limit.

As so defined, the area covered by the OSPAR Convention includes the internal waters and territorial seas of Ireland and the United Kingdom as well as the Irish Sea between them, but Article 1 does not indicate whether particular information is relevant to that maritime area.

171. Each of the second and third categories of Article 9(2) relates to "activities or measures." Neither of these terms is defined in Article 1 of the Convention, but it is clear from other parts of the OSPAR Convention (e.g., Article 2(1)(a)) that the term "measures" refers generically to regulatory initiatives by any part of the governmental apparatus of the Contracting Parties with respect to matters covered by the OSPAR Convention, while "activities" refers to the actions, whether emanating from or effected by governmental or non-governmental entities, that would be the object of the "measures."

172. In commenting on identical language in Article 2(a) of Directive 90/313, the ECJ in *Mecklenburg*,[122] remarked on "the term 'measures' as serving merely to make it clear that the acts governed by the directive included all forms of administrative activity."[123] Plainly, the inclusion of both "activities" and "measures" indicates that the drafters intended a regime in the second category covering "any available information" about a wide, rather than narrow, range of matters relating to the specific subject matter of each of those categories, but the Tribunal notes, once again, that the information must relate to the state of the maritime area.

173. The second category of Article 9(2) relates to two types of activities or measures. First, activities or measures that are already adversely affecting the maritime area and, second, activities or measures that are likely to affect it. The second type of activity or measure may be underway and already be affecting or likely to affect adversely the maritime area or it may not be underway, but if and when it is, it must be likely to affect adversely the maritime area if it is to fall within the second category. Thus the second category of Article 9(2) includes prospective activities and measures as well as activities and measures already underway.

174. Each of the three categories in Article 9(2) is cast in the broad terms that are consistent with the "any information" formula. As such, they might warrant an interpretation of inclusive causality. However, it is only the second category that contains an additional threshold of inclusion/exclusion that is manifestly designed by the drafters to be more restrictive than the first and third categories. While the scope of Article 9 covers *simpliciter* "any available information" "on the state of the maritime area" (first category) and "any available information" "on activities

---

122. *See supra* note 106.

123. *Id.*, at para. 20.

or measures introduced in accordance with the Convention" (third category), the second category of Article 9(2) qualifies the obligation to provide "any available information" on activities or measures "adversely affecting or likely to affect" the maritime area.

175. The adverb "adversely" qualifies both existing and prospective activities and measures and raises the threshold of inclusion, as does the adverb "likely." Even were the Tribunal to accept, *arguendo*, Ireland's submission of inclusive causality, the submission would founder on the adverb "adversely" and "likely." Had the adverbs "adversely" and "likely" not been inserted in the provision, the scope of that part of Article 9(2) would have included any present or prospective activity or measure having *any* effect on the maritime area and might, as a result, have indicated an intention of inclusive causality. By including those two adverbs, the drafters have excluded from the scope of the obligation of Article 9 current activities or measures that affected or were likely to affect the maritime area, but did not affect it *adversely* and prospective activities that were not likely to affect *adversely* the maritime area.

176. It may be that the object and purpose of this restrictive provision was based on a *de minimis* policy and was intended to preclude claims under Article 9 for available information about activities and measures that did not have adverse impacts on the maritime area. Alternatively, the restrictive character of the language may simply reflect a reluctance on the part of the Contracting Parties, at least at that stage, to undertake a broader obligation. In either case, the restrictive effect of the language in the second category is clear and is the standard which the Tribunal must apply.

177. The relevant parts of the *travaux préparatoires* show that Article 9(2) drew upon Directive 90/313. Article 2(a) of Directive 90/313, which speaks of "information relating to the environment," also establishes, as the criterion of inclusion, activities and measures "adversely affecting or likely so to affect . . . ."[124] The decision of the ECJ in *Mecklenburg* (para. 150 above), relied upon in this regard by Ireland, is not helpful. The Court was not there concerned with how the word "adversely" should be interpreted, but with how inclusively the term "information relating to the environment" should be construed.[125]

---

124. The Tribunal notes a minor discrepancy between the language of the Directive and that of the OSPAR Convention – namely, that the former includes the phrase "likely *so* to affect" (italics supplied) rather than "likely to affect." However, the drafting history in the record gives no indication that the word "so" was dropped with meaningful intent, and it is the Tribunal's view that the phrases were both intended to express a requirement of adverse effect of potential activities as well as current ones.

125. *Supra* note 106, at para. 6. Curiously, the adverb "adversely" appears to have been dropped in the *Umweltinformationsgesetz* of 8 July 1994 which transposed the Directive into German law.

178. In fact, the phrase "information relating to the environment" does not appear in Article 9(2) of the OSPAR Convention. Even if such a phrase did, it is doubtful if that would help Ireland, for it is far from clear that the 14 categories of redacted information identified by Ireland would fall within even that broader class. In any case, even if they did, the ultimate question is not the inclusiveness of the word "information" in Article 9(2), but the effect to be given to the additional and qualified threshold of adverse effect which is established by the second category of that provision.

179. In the opinion of the Tribunal, Ireland has failed to demonstrate that the 14 categories of redacted items in the PA and ADL Reports, insofar as they may be taken to be activities or measures with respect to the commissioning and operation of a MOX Plant at Sellafield, are "information . . . on the state of the maritime area" or, even if they were, are likely adversely to affect the maritime area.

180. Rather than engage the requirement of establishing an adverse effect, Ireland has focused its arguments on the questions of directness of the effect and whether or not the information considered as a whole was "environmental." To buttress its arguments, Ireland has sought to rely upon treaties that are as yet unratified and not in force as between the Parties or regional legislative initiatives that have not been finalized nor entered into force for the Parties. Although, it is arguable – but in the view of the Tribunal not conclusive – that Ireland's claim might have succeeded under some of these drafts, the Tribunal is not empowered to apply legally unperfected instruments. The OSPAR Convention does not adopt a lower threshold requiring no more than an activity or measure that "affects" rather than one that "affects adversely" the maritime area.

181. Ireland has also argued that Article 9(2) relates to any environmental information as such. But, wholly aside from the difficult question of whether the PA and ADL Reports dealt with environmental information (as opposed to information about economic justification), the words "environmental information" do not appear in Article 9(2) nor, indeed, in any part of Article 9. Even if such words did, it is doubtful that the 14 categories listed in paragraph 161 above would come within that class.

182. Hence the Tribunal finds that Ireland has not established that the class of redacted information that it seeks from the PA and ADL Reports under the second category of Article 9(2) falls under Article 9(2).

X. COSTS

183. The Rules enabled the Tribunal to award costs, and each Party argued for its costs.

184. The dispute is the first consideration of the OSPAR Convention. The Parties engaged the issues at a high level of dispassionate and professional competence. In all the circumstances, it is the Tribunal's decision to order that the costs of the Tribunal be equally divided between the Parties, and that there be no other order for costs.

XI. CONCLUSION

185. For the above reasons, the Tribunal

(i) by unanimous decision rejects the United Kingdom's request that the Tribunal find that it lacks jurisdiction over the dispute;

(ii) by unanimous decision rejects the United Kingdom's request that Ireland's claims are inadmissible;

(iii) by majority decision rejects the United Kingdom's submission that the implementation of Article 9(1) is assigned exclusively to the competent authorities in the United Kingdom and not to a tribunal established under the OSPAR Convention;

(iv) by majority decision finds that Ireland's claim for information does not fall within Article 9(2) of the OSPAR Convention;

(v) by majority decision finds that as a consequence, Ireland's claim – that the United Kingdom has breached its obligations under Article 9 of the OSPAR Convention, by refusing, on the basis of its understanding of the requirements of Article 9(3)(d), to make available information – does not arise; and

(vi) by unanimous decision decides that each Party will bear its own costs and an equal share of the costs of this arbitration.

AWARD OF THE ARBITRAL TRIBUNAL

FINAL AWARD

The Claims by Ireland are dismissed.

Professor W. Michael Reisman
Chairman

Gavan Griffith QC

Lord Mustill

ATTACHMENTS

Declaration of Professor W. Michael Reisman

Dissenting Opinion of Dr. Gavan Griffith QC

Declaration of Professor W. Michael Reisman

1. I do not concur in the majority's interpretation of Article 9(1) of the 1992 Convention for the Protection of the Marine Environment of the North-East Atlantic ("the OSPAR Convention").[1] In my opinion, Ireland's proposed interpretation of Article 9(1) should have been rejected.

2. The question here, as I understand it, is

   (i) whether Article 9(1) of the OSPAR Convention requires Contracting Parties to establish an internal regime that must make information – as defined by Article 9(2) and subject to the exceptions set out in Article 9(3) – available to persons requesting it, in the ways specifically prescribed by Article 9(1)

   or

   (ii) whether Article 9(1) simply requires Contracting Parties to make the information available.

   In the context of this case, the first of these possible interpretations would make the exclusive forum for disputes about *applications in specific cases* of the United Kingdom's obligations under this part of the OSPAR Convention the appropriate United Kingdom municipal law institutions. The only question for an international tribunal established under Article 32 of the OSPAR Convention with respect to complaints about the United Kingdom's Article 9(1) performance would be whether the United Kingdom had, in fact, established an internal regime that meets the requirements of Article 9(1).

3. The United Kingdom proposes the first of these two possible meanings while Ireland proposes the second. Everyone agrees that the answer to the question as to which meaning is the proper one is to be found in the text and context of the OSPAR Convention as interpreted in accordance with the Vienna Convention on the Law of Treaties ("Vienna Convention").[2]

---

1. Convention for the Protection of the Marine Environment of the North-East Atlantic, 22 September 1992, 32 ILM 1069 (1992).

2. Vienna Convention on the Law of Treaties, 23 May 1969, 1155 UNTS 331.

4. Article 9(1) of the OSPAR Convention provides:

> The Contracting Parties shall ensure that their competent authorities are required to make available the information described in paragraph 2 of this Article to any natural or legal person, in response to any reasonable request, without that person's having to prove an interest, without unreasonable charges, as soon as possible and at the latest within two months.

5. Ireland's proposed meaning would require deletion of a critical phrase in Article 9(1), which can be demonstrated by an overstrike of seven critical words.

> The Contracting Parties shall ~~ensure that their competent authorities are required to~~ make available the information described in paragraph 2 of this Article to any natural or legal person, in response to any reasonable request, without that person's having to prove an interest, without unreasonable charges, as soon as possible and at the latest within two months.

6. But the drafters of the OSPAR Convention included those seven words. In the *Anglo-Iranian Oil Company* case, the International Court of Justice ("ICJ") observed that the principle "that a legal text should be interpreted in such a way that a reason and a meaning can be attributed to every word in the text . . . should in general be applied when interpreting the text of a treaty."[3] The injunction is pertinent. The words "ensure that their competent authorities are required to," which Ireland's submission would require the Tribunal to ignore, make Article 9(1) an obligation to adjust domestic law in a prescribed way by providing for certain institutional recourses, for which specific criteria are provided. Article 9(1) is not expressed in terms to establish an obligation on the international plane to provide information, with the performance of that obligation in specific cases to be subject to the jurisdiction of a Tribunal established under Article 32.

7. This plain reading of Article 9(1) appears both to reflect its objects and purposes and to produce a reasonable and economic means for implementing Article 9(2) and (3) obligations. Indeed, it would be rather anomalous and duplicative for Article 9(1) to require Contracting Parties to ensure that their national competent authorities should do something and to prescribe how it should be done, yet then to assign the application role in specific cases to an international tribunal. The more plausible reading in terms of objects and purposes is that the implementation of the obligations and exceptions to those obligations for providing access to information was to be effected by those same national institutions.

---

3. *Anglo-Iranian Oil Co. (United Kingdom v. Iran)*, Preliminary Objection, Judgment of 22 July 1952, 1952 ICJ Rep. 93, at p. 105.

8. This reading is consistent with the achievement of other goals expressed in Article 9(1). Recall that Article 9(1) also requires that the internal procedures make the information available within a reasonable time and in any case in no more than two months. This demand for expedited decision is consistent with cognate municipal freedom of information actions, which usually involve current political issues, with respect to which delay about determining whether the information should be provided to the public would either deprive citizens of a right which the law seeks to protect or would suspend collective political action because the issue is *sub judice*. Such a delay could compromise the rights of other citizens to a timely decision. But the dispute resolution mechanism of Article 32 of the Convention is a slow and cumbersome one, which, under the best of circumstances, could not possibly be accomplished within two months.

9. If the Parties plainly intended to do something internally inconsistent or even, *arguendo*, producing an unreasonable result, an arbitrator might, strictures of the Vienna Convention notwithstanding, feel obliged to give effect to their intention. But such a hypothetical intention is far from plain and encounters textual and historical difficulties. As a textual matter, Article 9 is the only provision in the Convention which refers to another dispute resolution mechanism, in this instance a municipal one; this singularity hardly suggests that the obligations of Article 9 were being given the same treatment as the obligations of the other provisions of the Convention. As a historical matter, neither of the antecedent treaties that was incorporated in the OSPAR Convention, the Oslo Convention of 1972 and the Paris Convention of 1974,[4] included a provision comparable to Article 9. The Oslo Convention had no dispute resolution mechanism. The Paris Convention included arbitration provisions akin to Article 32 of the OSPAR Convention. The introduction into the OSPAR Convention of Article 9, which is thus unique among all the obligations imposed by the treaty in that it alone has its own dispute resolution mechanism, suggests that the Contracting Parties did not intend to subject those aspects of Article 9 that were assigned to a municipal remedy to the international provisions of Article 32.

10. The *travaux préparatoires* of Article 9(1) indicate that the language of what was to become Article 9(1) was adjusted to ensure that it was in conformity with EC Directive 90/313/EEC ("Directive 90/313") on Freedom of Access to Environmental Information.[5] This fact, which was not in contention, should have led the Tribunal, according to the United Kingdom, to read Article 9(1), in the light of

---

4. The Convention for the Prevention of Marine Pollution by Dumping from Ships and Aircraft (Oslo), 932 UNTS 3 (1972), and the Convention for the Prevention of Marine Pollution from Land-based Sources (Paris), 13 ILM 352 (1974).

5. Directive 90/313/EEC, 1990 OJ (L 158) 56.

the same objects and purposes as the Contracting Parties would appear to have been pursuing in Directive 90/313. As against this, Ireland referred to the fact that Directive 90/313 specifies that recourse would be had in domestic courts, while Article 9 does not do so. Article 4 of Directive 90/313 states:

> A person who considers that his request for information has been unreasonably refused or ignored, or has been inadequately answered by a public authority, may seek a judicial or administrative review of the decision in accordance with the relevant national legal system.

11. Ireland's observation is correct, as far as it goes, but it seems to me to be incomplete, for the comparison of Directive 90/313 with Article 9(1) underlines how closely linked is the obligation of Directive 90/313 and an exclusive municipal remedy. It is striking that the currently operative revision of Directive 90/313 of 2003,[6] which comes long after the OSPAR Convention, could have adopted an international remedy akin to Article 32 of the OSPAR Convention or, indeed, assigned decision competence to a standing instance within the European system. Instead, it reinforced and, indeed, elaborated the municipal remedial process.

12. The result of the interpretation of Article 9 that I believe is required is consistent with a not uncommon treaty practice in which states are obliged to make adjustments in domestic law and, to the extent that they do so appropriately, they have fulfilled their treaty obligations. The International Law Commission ("ILC") has had occasion to review this practice in the course of its work on International Liability for injurious consequences arising out of acts not prohibited by international law.[7] Article 5 of the ILC draft on this subject provides:

> States concerned shall take the necessary legislative, administrative or other action including the establishment of suitable monitoring mechanisms to implement the provisions of the present articles.[8]

13. In the third paragraph of its Commentary to this provision, the ILC said:

> To say that States must take the necessary measures does not mean that they must themselves get involved in operational issues relating to the

---

6. Directive 2003/4/EC on Public Access to Environmental Information and Repealing Council Directive 90/313/EEC, 2003 OJ (L 41).

7. *International Liability for injurious consequences arising out of acts not prohibited by international law (prevention of transboundary harm from hazardous activities)*, Report of the International Law Commission, 53rd Session, Supp. No. 10, UN Doc. A/56/10, 366 (2001).

8. *Id.*, at p. 398.

> activities to which article 1 applies. *Where these activities are conducted by private persons or enterprises, the obligation of the State is limited to establishing the appropriate regulatory framework and applying it in accordance with those articles. The application of that regulatory framework in the given case will then be a matter of ordinary administration or, in the case of disputes, for the relevant courts of tribunals*, aided by the principle of non-discrimination contained in article 15.[9]

14. In this regard, I find the recent decision of the ICJ in the *LaGrand* case,[10] construing the Vienna Convention on Consular Relations, instructive. Article 36(2) of the Vienna Convention on Consular Relations, like Article 9(1) of the OSPAR Convention, obliges States to ensure that their municipal laws enable an object of the Convention. In the former case, States must ensure that their municipal laws enable full effect to be given to the consular rights and obligations enumerated in Article 36(1); in the latter, States must "ensure that their competent authorities are required [by some municipal law mechanism] to make available the information" described in Article 9(2). In both cases, the only international claim that lies is that the respondent State failed to ensure that its municipal law was created or structured in such a way as to accomplish the objectives prescribed by the Convention. A direct claim for failure to accomplish those objectives in a specific case (provision of consular rights or certain information, respectively) does not lie because that is not how the specific obligation imposed by the relevant treaty provision is framed.

15. Thus, the ICJ said that if United States municipal law should continue to fail to enable full effect to be given to Article 36(1), thereby violating Article 36(2), with the result that a foreign national suffers prolonged detention or severe punishment, the United States must permit "review and reconsideration" under its municipal law. But because the treaty provision that creates the obligation, Article 36(2) of the Vienna Convention on Consular Relations, speaks of ensuring that a State's municipal law give full effect to the purposes for which the rights in Article 36(1) are intended, the Court observed that "[t]his obligation can be carried out in various ways. The choice of means must be left to the United States."[11]

16. Ireland contended that the United Kingdom's Department for the Environment, Transport and the Regions ("DETR") and its successor the Department of the Environment, Food and Rural Affairs ("DEFRA") were the competent authorities

---

9. *Id.*, at p. 399 (italics supplied).

10. *LaGrand (Germany v. United States of America)*, Judgment of 27 June 2001, 2001 ICJ Rep. 104.

11. *Id.*, at para. 125.

and that it requested the information from DEFRA and from the Foreign and Commonwealth Office without success. I would see the question of the identification of the competent authority as a matter to be resolved by the United Kingdom's law and review procedures. Where, as here, international law assigns a competence to national law, the assumption is that claims are to be pursued through the review mechanisms of that national system.

17. Ireland has also argued that, as a claimant, it is entitled to choose its forum. But this begs the question – whether cases of specific applications of Article 9(1) may be brought to a tribunal established under Article 32 – by assuming that the Convention provides for two fora: one under Article 9(1), the other under Article 32. In my view, questions of specific applications under Article 9(1) are assigned to the competent authorities within the United Kingdom. Hence there is no choice of forum.

18. The interpretation that appears correct to me does not mean that Article 9(1) is not subject to international standards. Although such a provision must allow a certain discretion or "margin of appreciation" as to its implementation to the Contracting Parties, the national arrangements must nonetheless meet whatever objective criteria are set out in the provision if they are not to be in breach of the Convention.

19. The objective criteria specified in Article 9(1) are that the information must be made available:

(i) to any natural or legal person;

(ii) in response to a reasonable request;

(iii) without the requester having to prove an interest;

(iv) without unreasonable charges; and

(v) as soon as possible but at the latest within two months.

20. Issues relating to alleged violations of these criteria would have been admissible under Article 32. In fact, Ireland does not claim that there are defects within the system that the United Kingdom had in place in its municipal law with respect to its obligations under Article 9(1), such that the United Kingdom's system falls below the standards required by the OSPAR Convention. It would be difficult to make such a claim, as Regulation 3 of the 1992 United Kingdom Regulations, the relevant legislative component in the United Kingdom, gives effect to Directive 90/313.

21. Ireland also sought to rely upon the fact that the OSPAR Convention has not been incorporated in United Kingdom law. In my view, the test of compliance at the national level for this sort of provision is substantive and not formal. In requiring that "Contracting Parties *ensure* . . .", Article 9(1) does not, by its terms, require a party to enact or re-enact dedicated confirmatory laws, if its domestic competent authorities are already legally bound under municipal law to make the relevant information available in ways that meet the criteria set out in that provision.

W. Michael Reisman
2 July 2003

Dissenting Opinion of Gavan Griffith QC

I here express my reasons for my disagreement with Parts VII and IX of the Majority Opinion and my dissent from the majority's decision to dismiss Ireland's claims.

For the reason that the majority has determined in Part IX that the whole of the redacted material in the PA and ADL Reports is not information within Article 9(2) I have joined in signing the Final Award as dispositive of the dispute.

Applicable Law – Part VII in the Majority Opinion

1. I disagree with the reasons of the restrictive interpretation of applicable law adopted by the majority and its rejection of the normative value of various international instruments invoked by Ireland in support of its position.

2. I regard other international legal sources as having direct relevance to the construction of this arbitration for several reasons –

    (1) As constituted under Article 32 of the OSPAR Convention,[1] the Tribunal is instructed to take into account relevant rules of international law in terms that: *"the arbitral tribunal shall decide according to the rules of international law and, in particular, those of the Convention."* The content of this mandate must be informed by Article 38 of the Statute of the International Court of Justice ("ICJ") as the authoritative and orthodox catalogue of sources of international law. The Tribunal cannot be confined to international conventional law or the language of the OSPAR Convention exclusively. Customary rules and general principles of law, as embraced by Article 38, are as much included in the general reference to "*rules of international law*" in Article 32(6)(a) as are treaties. Contrary to the majority (para. 92[2]), the Tribunal's mandate in this regard is not confined by the absence in Article 9(1) and 9(2) of specific reference to any other body of substantive conventional law.

---

1. Convention for the Protection of the Marine Environment of the North-East Atlantic, 22 September 1992, 32 ILM 1069 (1992).

2. This and all further paragraph references, unless otherwise indicated, are to the Majority Opinion.

(2) As an interpreting agency, the Tribunal is guided by Article 31(3)(c) of the Vienna Convention on the Law of Treaties ("Vienna Convention") which refers to rules of international law that correspond to the catalogue of sources in Article 38 of the ICJ Statute. Such rules must be relevant (i.e. concern the subject matter in question) and applicable at least to Ireland and the United Kingdom.

(3) Further and in any event, the language of Article 9(3) specifically directs the Tribunal to take account of international regulations.

3. As it is common ground between Ireland and the United Kingdom ("the Parties") that the reference to international rules and regulations in the OSPAR Convention cannot extend the competence of the Tribunal to the consideration of obligations the Parties might have under other instruments, the discussion of the majority on this issue in paragraph 85 is merely a confirmatory statement of an agreed position.

4. Further, I accept that the OSPAR Convention stands as a *lex specialis* between the Parties to be interpreted within the general context of other relevant rules and principles of international law. Its drafters plainly perceived the OSPAR Convention as an integral part of a matrix of international instruments directed to environmental protection. The Preamble affirms that the Convention has been drafted to be consistent with customary and conventional international law. Further, its provisions repeatedly and explicitly require close consideration of international legal sources, including –

– Article 1 under paragraphs –

(a) (*"to the extent recognised by international law"*);

(g) (i) (*"other applicable international law"*);

(iii) (*"other relevant international law"*); and

(h) (i) (*"applicable international law"*);

– Article 7 (*"other international conventions"*);

– Article 10(1)(c) of Annex II (*"to the extent recognised by international law"*);

- Article 9(2) of Annex III (*"entitled under international law"*); and
- Article 3(1)(b)(ii) of Annex V (*"consistent with international law"*).

5. Hence, the explicit mandate of the Tribunal in interpreting Article 9 is to apply the OSPAR Convention as a *lex specialis* between the Parties consistently with "international law" broadly defined, and not confined merely to treaty and conventional law in force binding on the Parties (*cf.* para. 105).

6. To my mind the incantation by the majority, in paragraph 180, that *"the Tribunal is not empowered to apply legally unperfected instruments"* does not excuse the Tribunal from the examination of the extent to which the points of intentional reference invoked by Ireland inform the construction of Article 9(2).

7. In this regard, I depart from the majority's rejection of the normative value and applicability of the various international instruments invoked by Ireland, and in particular its rejection of the relevance of the Aarhus Convention[3] and EC legislative proposals to inform the meaning of Article 9(2) on the grounds that these instruments are not law and that the Tribunal is not instructed to apply law *in statu nascendi* (paras. 99 to 106).

8. I support my position by analysis of the status of these instruments.

Aarhus Convention

9. Since neither of the Parties to this dispute has ratified it, I accept that the Aarhus Convention is not *ex facie* an instance of binding international law under Article 31(3)(c) of the Vienna Convention. I also accept the United Kingdom's contention that the question of interpretation of the Aarhus Convention lies outside the scope of the competence of this Tribunal.[4]

10. However, it does not follow that the Aarhus Convention cannot inform the issues of construction of Article 9 that arise in this dispute. To the contrary, it is my opinion that the Aarhus Convention, to which each of the United Kingdom and Ireland is a signatory, does have a relevant normative and evidentiary value that

3. Convention on Access to Information, Public Participation in Decision-making and Access to Justice in Environmental Matters, 25 June 1998, 38 ILM 517 (1999).

4. United Kingdom's Counter-Memorial, para. 4.13. Counsel for the United Kingdom, Transcript, Day 2 Proceedings, p. 52 and p. 87.

is not denied merely because neither the United Kingdom nor Ireland has yet ratified it.

11. The Aarhus Convention entered into force on 30 October 2001 (that happens also to be the date of constitution of the Tribunal). Some 22 of the 39 signatory States have subsequently ratified it. Hence, as a *lex lata*, the majority is incorrect to discard the Aarhus Convention as merely *"'almost' law"* (para. 99) or *"material that has not yet become law"* (para. 103).

12. Although certain provisions of the Aarhus Convention may be recognised as reflecting or codifying customary practice and general principles of international law that are binding on the Parties, the Tribunal has no competence to pronounce on the customary nature of the provisions of the Aarhus Convention.

13. However, and at the least, Article 18 of the Vienna Convention applies to require the United Kingdom as a signatory State to the Aarhus Convention, to refrain from acts that would defeat its objects and purposes. Hence, to a limited extent it may be said that the Vienna Convention has the effect that the United Kingdom is bound by its object and purpose pending ratification.

14. Speaking generally, the discernment of the object and purpose of a treaty is a flexible and abstract process. As well as the text of a treaty itself,[5] interpreting bodies may derive understanding from various sources, including the preambular provisions and *travaux préparatoires*,[6] titles of treaties[7] and even their spirit.[8] Although some writers have proposed that the object and purpose may be interpreted as *"a normative element beyond the rules laid down in a treaty"*,[9] for present purposes it is unnecessary to advance to such further reaches of international sources of meaning and approach.

15. Independently of the Vienna Convention, the recent *dictum* of the ICJ in *Qatar v. Bahrain* accepted the principle that unratified treaties may possess an

---

5. United Kingdom's Counter-Memorial, para. 273.

6. *See*, for instance, *Reservations to the Convention on the Prevention and Punishment of the Crime of Genocide*, Advisory Opinion of 28 May 1951, 1951 ICJ Rep. 15, at p. 23.

7. Diversion of Water from the River Meuse, 1937 PCIJ (Ser. A) No. 70, at 21.

8. *Case Concerning Military and Paramilitary Activities in and Against Nicaragua (Nicaragua v. United States of America)*, Merits, Judgment of 27 June 1986, 1986 ICJ Rep. 14, at para. 275.

9. Willem Riphagen, State Responsibility: New Theories of Obligations in Interstate Relations, *in* THE STRUCTURE AND PROCESS OF INTERNATIONAL LAW 581 et seq., and at 601 (R. St.J. McDonald and D.M. Johnson, eds., 1983). *See also* the *Nicaragua* case, *supra* fn. 8, paras. 270-182.

evidentiary value that help establish and identify the views and intentions of signatories, in terms that –

> *signed but unratified treaties may constitute an accurate expression of the understanding of the parties at the time of signature.*[10]

16. This principle has apparent application to the process of determining whether Ireland is correct to contend that the Aarhus Convention may be invoked to inform the proper construction of Article 9(2). At the least, it enables the Tribunal to regard the Aarhus Convention as evidence of the views of the United Kingdom and Ireland on the scope of the definition of environmental information.[11]

17. Even though this cannot be used as a self-standing legal argument, in fact the United Kingdom has maintained its intention to be bound by, and to implement, the obligations of the Aarhus Convention.[12] Most recently, in its Proposal for a Revised Regime for Public Access to Environmental Information,[13] the Department of the Environment, Food and Rural Affairs ("DEFRA") reiterated that *"the UK is committed to ratifying the Aarhus Convention as soon as possible"*.[14] After noting that *"the Freedom of Information Act 2000 contains a power to enable Regulations to be made to bring the existing regime into line with the Aarhus regime ahead of action at the Community level (such as this proposal)"*,[15] DEFRA went on to identify a number of specific examples confirming that the

---

10. *Maritime Delimitation and Territorial Questions between Qatar and Bahrain (Qatar v. Bahrain)*, Judgment of 16 March 2001, 2001 ICJ Rep. 40, at para. 89.

11. In its Judgment, the ICJ held that "the Anglo-Ottoman Convention does represent evidence of the views of Great Britain and the Ottoman Empire as to the factual extent of the authority of theAl-Thani ruler in Qatar up to 1913". Further, the ICJ gave special consideration to the unambiguous language of the 1913 Treaty and the fact that Article 11 of the said treaty was later incorporated into the Anglo-Ottoman Treaty of 1914. *Id.*, at paras. 89-91.

12. On 15 July 2002, the United Kingdom Department of Environment, Food and Rural Affairs ("DEFRA") launched a "public consultation period for new Environmental Information Regulations, illustrating the government's commitment to freedom of information and to greater openness and transparency . . . . The new regulations are a step towards the full implementation of the Freedom of Information Act 2000 and will enable the UK to fulfil its obligations under the UNECE (United Nations Economic Commission for Europe) 'Aarhus Convention'". *See* DEFRA's website at www.defra.gov.uk.

13. Public Access to Environmental Information. Proposals for a Revised Regime. Regulatory Impact Assessment. Proposal for a Directive of the European Parliament and of the Council on Public Access to Environmental Information [COM(2000) 402 Final]. This document can be found at DEFRA's website at www.defra.gov.uk.

14. *Id.*, para. 4.

15. *Id.*, para. 7.

United Kingdom is already incorporating provisions of the Aarhus Convention into its domestic legislation.[16]

18. Hence, although the formal act of ratification that would establish on the international plane the consent of the United Kingdom to be bound by the Aarhus Convention has not yet occurred, the United Kingdom's intention to treat the Aarhus Convention as a binding instrument is unequivocally confirmed.

19. Contrary to the majority I conclude that the Aarhus Convention falls within the definition of applicable law and Article 31(3)(c) of the Vienna Convention as a legal source that possesses some normative and evidentiary value to the extent that regard may be had to it to inform and confirm the content of the definition of information contained in Article 9(2) of the OSPAR Convention.

International Practice

20. In its Memorial, Ireland invited the Tribunal to interpret the definitions of Article 9(2) in accordance with international law and practice.[17]

21. The stated grounds of the majority (paras. 100 and 101) in refusing to take into account various examples from international practice invoked by Ireland as informing the meaning of information are that the Tribunal was not requested by the Parties to apply evolving law and that a similar provision cannot be found in the language of the OSPAR Convention. That much is accepted. Nonetheless, in my opinion international practice nonetheless is a relevant matter for consideration.

22. Again the Vienna Convention is invoked. The relevance of the EC and United Kingdom legislative proposals is founded on Article 32(2)(b), which directs consideration of *"any subsequent practice in the application of the treaty which establishes the agreement of the parties regarding its interpretation"*. Relevantly the reference here is not to *law* but to the *subsequent practice* in the application.

23. Although the particular curial references relied upon by Ireland may be relevant for interpreting EC Directive 90/313/EEC ("Directive 90/313"),[18] their application to the OSPAR Convention is not so obvious. However, it is the common ground between the Parties that the definitions of the OSPAR Convention closely

16. *Id.*, paras. 9, 12, 16, and others.

17. Ireland's Memorial, paras. 100-101.

18. Directive 90/313/EEC, 1990 OJ (L 158) 56.

follow the language of this Directive, on which it appears to be based.[19] Further the drafters of Directive 90/313 and the OSPAR Convention were much the same persons representative of the same States and international organisations. To my mind, ordinary principles of comity and interpretation may here be invoked to suggest that the same State parties broadly may be assumed to understand similarly or identically worded obligations in the same way.

24. In this regard, the subsequent practice of the European Union as a principal party to the OSPAR Convention,[20] also may be invoked as directly relevant to the interpretation of its provisions.[21] There is an emerging recognition that the practice of international organisations and State parties to an international agreement may inform the interpretation of that agreement: for example, in the 1998 *Fisheries Jurisdiction (Spain v. Canada)* case the ICJ examined the meaning of *"conservation and management measures"* by referring to the subsequent practice of States, the European Union (*e.g.* EC Regulations) and North-West Atlantic Fisheries Organization ("the NAFO").[22]

25. This is not to say that the construction of Article 9 is in any relevant sense to be driven by regard to subsequent practice. The Tribunal here should be conser-

---

19. *See*, for instance, Counsel for Ireland, Transcript, Day 1 Proceedings, p. 26: "Directive 90/313 whose provisions are to all intents and purposes identical to those of Article 9"; Counsel for the United Kingdom, Transcript, Day 2 Proceedings, p. 78: "The travaux confirm that the wording of Articles 9(1) and 9(3) were specifically amended in order to secure conformity with Articles 3(1) and (2) of Directive 90/313/EEC."

20. The OSPAR Convention has been signed by the representatives of the EC Commission on behalf of the European Union. *See* http://www.ospar.org/eng/.

21. It should be noted that the 1969 Vienna Convention on the Law of Treaties does not include international organisations in the definition of a party to an international agreement (*see* Article 1(g)). However, there are claims as to the emergence of customary rules on this issue. The capacity of international organisations to conclude international agreements and participate in the interpretation and modification of their terms has been recognised by the International Law Commission itself. *See*, for instance, the Preamble to the 1986 Convention on the Law of Treaties Between States and International Organisations or Between International Organisations. Article 31 of the 1986 Convention is identical to Article 31 of the 1969 Vienna Convention on the Law of Treaties.

22. ". . . The same usage is to be found in the practice of States. Typically, in their enactments and administrative acts, States describe such measures by reference to such criteria as . . . (see, among very many examples, Algerian Legislative Decree No. 94-13 of 28 May 1994 . . . as well as, for the European Union, the basic texts formed by Regulation (EEC) No. 3760/92 of 20 December 1992, establishing a Community system for fisheries and aquaculture, and Regulation (EC) No. 894/97 of 29 April 1997, laying down certain technical measures for the conservation of fisheries resources. For NAFO practice, see its document entitled Conservation and Enforcement Measures (NAFO/FC/Doc. 96/1)). International law thus characterizes 'conservation and management measures' by reference to factual and scientific criteria." *See Fisheries Jurisdiction (Spain v. Canada)*, Judgment of 4 December 1998, 1999 ICJ Rep., para. 70.

vative in this aspect of interpretation, and give due weight to the consideration that the United Kingdom has not ratified the Aarhus Convention and that neither it nor the OSPAR Convention is an EU instrument. In my opinion, the Vienna Convention nonetheless may enable the EC legislative proposals and the 2000 DEFRA Proposals referred to by Ireland to be taken into account by this Tribunal so as to inform and confirm the proper construction of Article 9(2). At the least, these matters may be said to reflect subsequent interpretation by the Parties to the OSPAR Convention of a definition nearly identical to Article 9(2). In the context that a more consistent interpretation appears reasonably to be open, at the least such comparisons invite reflection as to whether the confined construction applied by the majority is correct on having regard to all relevant circumstances.

26. Further, I contest the failure of the majority to refer to, or to take into account, the adoption on 28 January 2003 of Council Directive 2003/4/EC ("Directive 2003/4") on public access to environmental information, in substitution for Directive 90/313.[23] Directive 2003/4 entered into force on 14 February 2003 and is required to be fully implemented by EU Member States by 15 February 2005. To my mind this instrument is a significant development in EU environmental legislation promulgated after the oral hearings. The majority does not consider it, presumably on the ground that as an EC instrument issued subsequent to the hearing it has no relevance to the issues of construction of Article 9(2).

27. I disagree with such a strict temporal approach. It must be a relevant matter for consideration that during the pendency of this Tribunal's award relevant EC legislative proposals have been transformed into a *lex lata* by Directive 2003/4.

28. Hence, even if the attribution of normative value to the Aarhus Convention or EC legislative practice is contested, Directive 2003/4 now independently constitutes a relevant international regulation that provides for a broad definition of information consistent with the meaning of Article 9(2) advanced by Ireland to apply as a binding obligation upon the United Kingdom.

29. As an instrument in force, it is my opinion that Directive 2003/4 must be accepted as material relevant to the determination of the meaning of "information" in Article 9(2). On its face Directive 2003/4 favours Ireland's interpretation of the definition of Article 9(2) as including commercial data and economic analyses, and it explicitly applies like definitions to those of the Aarhus

---

23. Directive 2003/4/EC on Public Access to Environmental Information and Repealing Council Directive 90/313/EEC, 2003 OJ (L 41).

Convention as relevantly binding on both the United Kingdom and Ireland.[24] Whether or not this is so, the point here made in consideration of the applicable law issue is that, at the least, it is necessary to consider the position of Directive 2003/4.

30. In this context, I regard the *Gabcíkovo-Nagymaros* case, relied upon by Ireland[25] and rejected by the majority as irrelevant in paragraph 101, as confirming these conclusions. There the ICJ observed that –

> *. . . new norms and standards have been developed, set forth in a great number of instruments during the last two decades. Such new norms have to be taken into consideration, and such new standards given proper weight, not only when States contemplate new activities but also when continuing with activities begun in the past.*[26]

31. Contrary to the majority (para. 103), I conclude that such new *law* here may be found in a number of instruments, and, in particular, in the Aarhus Convention and Directive 2003/4.

32. In summary, on the issue of applicable law I agree with the majority's statement (para. 103) that an international tribunal *"must certainly engage in actualisation or contemporization when construing an international instrument that was concluded in an earlier period"*. My criticism is that the majority adopts an impermissibly restrictive view in confining the applicable law to positive *lex scripta* by failing to take account of relevant, and to my mind applicable, existing international obligations of the Parties.

33. On the other hand, the relevance or permissive force of such other instruments must not be overstated to divert the enquiry from its objective to ascertain the applicable interpretation of the terms of Article 9. My departure from the majority is their refusal to have any regard to such other instruments.

---

24. *See*, in particular, para. 5 of Directive 2003/4 which reaffirms that it is meant to bring EU legislation in accord with the Aarhus Convention.

25. Ireland's Reply, pp. 15-16, para. 102.

26. *Case Concerning the Gabcíkovo-Nagymaros Project (Hungary v. Slovakia)*, Judgment of 25 September 1997, 1997 ICJ Rep., at p. 7, para. 140.

Findings with Respect to Article 9(2) – Part IX Section 2 of the Majority Opinion

Second Category of Information

34. The finding of the majority in rejection of the application of the second category of information under Article 9(2) is expressed as predicated on the position that –

(1) As for the first category, the second category also is limited to information *"on the state of the maritime area"* (paras. 163, 168 and 179), and

(2) the onus is on Ireland to establish that the MOX fuel production is an activity which is likely adversely to affect the maritime area, which Ireland has failed to demonstrate (*e.g.*, paras. 179 and the final 185).

I disagree with both conclusions.

On the State of the Maritime Area

35. On the first issue, I reject the majority's generalised interpretation of the definition of the second and third categories of information of Article 9(2) (advanced in paras. 163 to 168 of the Majority Opinion) of applying as an over-arching requirement that the information for each of the three categories be confined to information *"on the state of the maritime area"*.

36. The second and third categories specifically and relevantly are differently defined by reference to information *"on activities or measures adversely affecting or likely to affect"* the maritime area and *"activities or measures introduced in accordance with the Convention"* respectively. As categories separate from the first category, it is no part of the definition of either of these two later categories that they be confined to information *"on the state of the maritime area"*. Indeed, the third category is not defined by reference to the maritime area at all, rather that it concern *"activities or measures introduced in accordance with the Convention"*.

37. To make my objection plain, the majority in paragraph 163 asserts as a *"manifest"* conclusion that none of the 14 categories of information identified by them in the Majority Opinion may *"plausibly be characterized"* as *"information . . . on the state of the maritime area"* in terms that –

> *Article 9(2), whose chapeau is 'Access to Information,' establishes the scope of information to which, subject to specific enumerated rights of refusal in Article 9(3), the obligation in Article 9(1) relates. The scope of the information in the provision is not environmental, in general, but, in keeping with the focus of the OSPAR Convention, 'the state of the maritime area.' It is manifest to the Tribunal that none of the above 14 categories in Ireland's list can plausibly be characterized as 'information . . . on the state of the maritime area.'*

And in paragraph 168 that –

> *The point of emphasis, however, is that it is 'information' about the state of the maritime area. The three categories of 'information . . . on the state of the maritime area' in Article 9(2) are*
>
> *(i) 'any available information' on 'the state of the maritime area,'*
>
> *(ii) 'any available information' on 'activities or measures adversely affecting or likely to affect . . . the maritime area,'*
>
> *(iii) 'any available information' on 'activities or measures introduced in accordance with the Convention.'*

38. To my mind, it is more the case that it is the majority's error that here becomes manifest by reason of its misstatement of the terms of Article 9(2). As a matter of unambiguous grammatical construction, the expression of the second category of information is incapable of being confined to *"information . . . on the state of the maritime area"* in the same terms as the first category of information (not invoked by Ireland). It could not be more clearly expressed that it is no part of the definition of the second or third categories of information to requires attachment of the information as *"on the state of the maritime area"*. For this reason the majority's conclusion (in para. 163) that it could have rested *"its decision on the fact that none of the material in the 14 categories falls within the definition of 'information'"* cannot be sustained by the text of Article 9(2).

Wider Issues

39. I now turn to identify the other errors in the determination of the majority, including onus.

40. It is accepted that the Tribunal's task is not to issue an advisory opinion as to the abstract meaning of Article 9(2). To this end, Ireland's request for relief is sufficiently specific and directed to the disclosure of the 14 categories of information redacted from the PA and ADL Reports ("the Reports") as the legal remedy sought in these proceedings.

41. However, the scope of issues submitted by the Parties is not confined to the simplistic application of the definition under review to the 14 categories of redacted items. The Parties have called upon the Tribunal to interpret the extent and meaning of the definition of Article 9(2), with particular emphasis on the second category of information, and to establish whether a sufficient link exists between the Reports and measures or activities. Amongst others, the Parties disagree on whether the commercial nature of the Reports precludes them from falling within the scope of the definition. Hence, one of the main contentious issues between the Parties is the extent and inclusiveness of the definition.

42. In finding that the primary task of the Tribunal is to examine whether any of the 14 categories of redacted information fall within the definition of Article 9(2), the majority (para. 161) explicitly refuses to consider these wider issues raised for determination. Its approach suggests that the Reports ought to be dissected into separate pieces of information, each of which is to be tested against the definition.

43. Since the majority concludes (para. 179) that there is no activity that has the potential adversely to affect the marine environment, it does not go into the details of the adopted line of analysis. However, had the majority reached the contrary conclusion at this point, on its approach it would have been required then to engage in the examination and characterisation of each redacted item in its consideration of Article 9(2). At this secondary level such an analysis could not be limited to the categories of redactions in disregard of their contents. It then could not be established whether a link existed between a specific item and a hazardous activity without studying and extrapolating its exact content. Under such circumstances, the separation of form from content, as the majority suggests in paragraph 161, would have been unattainable. In turn, that might have lead to finding some sections of the Reports were undisclosable under Article 9(2) due to their failure to satisfy the necessary requirements of the assumed definition.

44. I reject this unnaturally confined approach of the majority. The correct position appears to me to be that the exercise of interpretation under Article 9(2) must engage the Tribunal in identifying whether the Reports as a whole *in principle* fall within the scope of the definition. At this level, the Tribunal's main task must be to elucidate and clarify the meaning of the terms used in Article 9(2) and to apply the established meaning to the entire Reports.

45. To this end, the regime for the disclosure of specific items is controlled by Article 9(1) to impose a wide obligation on the Contracting Parties to ensure that information is provided. Article 9(3) then lists specific exceptions to that obligation. Article 9(2) merely supports the definition of what is information within Article 9(1). In other words, once established that the information contained in each

Report is, in principle, within Article 9(2), the entire Reports have to be made available under the terms of Article 9(1) except as to parts protected as excepted matter under Article 9(3). There appears no room for a further analysis of redactions, category by category, in the Article 9(2) exercise in the manner summarily engaged by the majority.

46. My conclusion that this inclusive all-in or all-out approach under Article 9(1) and (2) is mandated as a matter of construction also appears to me to be consistent with international and also United Kingdom domestic jurisprudence.

47. In the *Mecklenburg* case the European Court of Justice ("ECJ") applied the definition of information contained in Directive 90/313 to a statement of views put forward by a countryside protection authority. In testing the statement against the definition the ECJ concluded that –

> *In order to constitute 'information relating to the environment for the purposes of the directive', it is sufficient for the statement of views put forward by an authority, such as the statement concerned in the main proceedings, to be an act capable of adversely affecting or protecting the state of one of the sectors of the environment covered by the directive. That is the case, as the referring court mentioned, where the statement of views is capable of influencing the outcome of the development consent proceedings as regards interests pertaining to the protection of the environment.*[27]

48. In finding that the statement as a whole was to be characterised as a measure, the ECJ did not suggest any analysis of specific provisions or parts of the statement in light of the Directive. Instead, the ECJ found that the appropriate test at this level is confined to the establishment of the existence of a causal link between an act and a hazardous activity. I see no basis for a different approach to characterisation under Article 9.

49. Further, in *R. v. Secretary of State & Ors ex p. Alliance against the Birmingham Northern Relief Road & Ors*,[28] cited by Ireland,[29] Sullivan J, sitting in the Queen's Bench Division of the High Court, held that the 1992 Environmental Regulations (implementing Directive 90/313) embraced, within the meaning of *"information relating to the environment"* an agreement for the construction of a toll motorway. Relevantly, he applied, as the appropriate approach to con-

---

27. C-321/96, *Mecklenburg v. Kreis Pinneberg – Der Landrat*, [1999] 2 CMLR 418, 435, at para. 21.

28. [1999] Envtl. L. Rev. 447, 470.

29. Counsel for Ireland, Transcript, Day 1 Proceedings, pp. 43-44.

struction, a wide view of the definition; and left it for the specific regulations protecting the release of commercially confidential information to limit the particular disclosure, in terms –

> *The fact that the Agreement can be described as a 'commercial document' does not mean that it does not contain information which related to the environment. It simply means that if such information is contained in the Agreement it may fall within one of the exceptions in regulation 4 . . . .*
>
> *The definition of 'information relating to the environment' in Article 2 of the Directive (90/313) is very broad, in my view deliberately so, and this broad definition has been carried through in to the Regulations . . . . The fact that . . . regulation 2 may cover a large range of documentation is not a valid argument for a narrow interpretation.*

50. I accept the analysis of Sullivan J is as apt also to describe the regime of Article 9 as for Directive 90/313 and the 1992 Environment Regulations. In pursuit of its policy for disclosure, each adopts a deliberately broad definition of information that implicitly may include commercial and confidential information, and then provides for extensive and defined categories of exceptions from that broad category. In the case of Article 9, the detailed and comprehensive regime for exceptions from the wide reach of the definition of environmental information falling within Article 9(1) and (2) is furnished by Article 9(3).

51. In this context, it appears to me that there is no scope for the introduction of an intervening limitation (as has been applied by the majority) to exclude categories of admittedly commercial information from the embrace of the general definition of information at the level of the definitions of Article 9(2). The scheme of both Directive 90/313 and Article 9 (and, if it matters, the applicable 1992 Environment Regulations giving effect in the United Kingdom to both Directive 90/313 and also the obligations of Article 9) is as stated by Sullivan J. Information in these excepted categories falls to be determined at the second level of the application of the excepting provisions to such general disclosure. In the case of Article 9, this work is done by Article 9(3), much as for the 1992 Regulations the regime for exception is provided by Regulation 4.

52. In my opinion, the majority is in error in applying its subjective approach that Article 9 could not have intended the disclosure of obviously commercial information at the level of the threshold definitions of information in Article 9(2) rather than leaving the issue to be resolved under the next level of the comprehensive scheme of exceptions under Article 9(3). The majority should have deferred to this plain definitional structure, and left the exceptions from disclosure to be determined at the level of Article 9(3).

53. My further point of criticism is that the majority focuses on the second category of information covered by Article 9(2) as the critical category indicated by the Parties.[30] In so doing the majority does not pay sufficient attention to the Tribunal's *Decision No. 5* issued 26 January 2003 requesting the views of the Parties set out in paragraph 76 of the Majority Opinion.

54. This request by the Tribunal was directed to clarify the Parties' positions on the relevance and meaning of the third category of information defined by reference to *"activities or measures introduced in accordance with the Convention"*. The Majority Opinion merely makes passing references to this issue without responding meaningfully to the responses advanced by the Parties.

55. In summary, I consider that the task of the Tribunal was to interpret the entire Article 9(2) and consider its definitions of the second and third categories of information to the contentious issues of –

(1) identifying an activity or measure which adversely affects or is likely to affect the marine environment of the Irish Sea;

(2) establishing whether the link exists between such an activity or measure and the PA and ADL Reports; and

(3) examining whether the information contained in the Reports may simultaneously fall under both the second and third category of information of Article 9(2).

56. At the outset of the hearing, the primary basis for Ireland's claims for the disclosure of the information redacted from the PA and ADL Reports was premised on the grounds that Ireland must be in a position to assess likely impacts on the maritime area, and should be enabled itself to engage in a process of assessing objectively the justification of the MOX Plant.[31] Ireland contended that the data contained in the PA and ADL Reports must constitute information on activities or measures likely to affect the maritime area[32] because (as the

---

30. *See* Ireland's Memorial, para. 96. *See also* United Kingdom's Counter-Memorial, para. 4.2; Counsel for Ireland, Transcript, Day 1 Proceedings, p. 50; Counsel for the United Kingdom, Transcript, Day 2 Proceedings, p. 86.

31. Counsel for Ireland, Transcript, Day 1 Proceedings, pp. 5-11 and 26.

32. In Ireland's words, ". . . [I]t is self-evident that the information in both Reports constitutes information 'on activities . . . adversely affecting or likely to affect [the maritime area]' within the meaning of Article 9(2) . . . . MOX production is an activity which will inevitably and certainly affect the maritime area, including Ireland's waters. It will do so principally in three ways: (1) routine (intentional) discharges from MOX; (2) routine (intentional) discharges from THORP, due to the intensification of

United Kingdom recognised) the MOX fuel production would have an adverse impact on the maritime area covered by the Convention.[33]

57. The United Kingdom answered that the relevant question was not whether the MOX production could affect the maritime area and more whether the information requested is information on activities or measures adversely affecting or likely to affect the maritime area.[34] The United Kingdom identified the relevant activity as *"the discharging of radioactive elements from the MOX plant into the maritime area, and not the wider activity of the MOX plant which in many of its aspects has nothing whatsoever to do with the maritime area"*.[35] Therefore, the information was not *"directly and proximately related to . . . activities or measures adversely affecting or likely to affect the maritime area"*.[36]

58. The United Kingdom also contested the fact that radioactive discharges from the MOX Plant were of a potentially adverse character.[37]

59. The Parties diverged on the threshold question whether the commercial nature of the Reports precluded them falling within the scope of Article 9(2). Although it obviously related to commercial activities, Ireland contended that the information

---

activity aimed at producing materials for the MOX plant; (3) discharges from possible accidents or terrorist attacks, either from the MOX plant itself or from transports of radioactive waste to, or MOX from, the plant." *See* Ireland's Memorial, para. 96. *See also* United Kingdom's Counter-Memorial, para. 4.2; Counsel for Ireland, Transcript, Day 1 Proceedings, p. 48.

33. Ireland's Memorial, para. 97. In support of this argument Ireland refers to the Sintra Ministerial Statement of 1998, in which the UK has itself recognised the long-term damage done to the marine environment by radioactive discharges, and has undertaken to reduce background radiation to "close to zero" by 2020. *Ibid. See also* Ireland's Memorial, Annex 5, referring to DEFRA Decision of October 2001. *See also* Ireland's Reply, para. 12; Counsel for Ireland, Transcript, Day 1 Proceedings, pp. 46-48.

34. United Kingdom's Counter-Memorial, para. 4.3.

35. Counsel for the United Kingdom, Transcript, Day 2 Proceedings, p. 86.

36. United Kingdom's Counter-Memorial, paras. 4.8-4.9.

37. Counsel for the United Kingdom, Transcript, Day 2 Proceedings, pp. 62-63.

also directly affected the environment,[38] and that, as such, the fact that it also was of a commercial character was irrelevant.[39]

60. The United Kingdom asserted that the Reports' plainly commercial nature was determinative, and that, once it had been concluded in the determination process that the balance was broadly neutral, no issue of information under Article 9(2) could arise with respect to the characterisation of this "commercial information". It contended that the information then ceased to be capable of being regarded relevant to the environment because –

> *. . . given [the Government's] conclusion on environmental and other issues, that the balance is broadly neutral, the draft decision then went on to consider the economic case concluded that there was a case for approval. This is the point in the stage of consultations at which consideration of the environmental issues was concluded and from this point onwards, essentially, the issues being considered are no longer environmental, . . . the issues considered hereafter were the commercial arguments for and against the plant or the process.*[40]

61. On this first issue of what constitutes a harmful activity to which the Reports may be related, I agree with the United Kingdom's characterisation of the relevant activity as *"the discharging of radioactive elements from the MOX plant into the maritime area"*.[41]

Adverse Effect

62. It follows that the second issue for determination is whether this activity may be characterised as having the potential adversely to affect the marine environment of the Irish Sea.

---

38. Counsel for Ireland, Transcript, Day 1 Proceedings, p. 48. *See also* para. 66 of the Written Outline of submissions on behalf of Ireland on file at the Permanent Court of Arbitration (Professor Sands): "the information relates to commercial activity, but it is (presumably) not in dispute that the consequences of the activity may be harmful to the environment".

39. Counsel for Ireland, Transcript, Day 1 Proceedings, p. 48. In support of this statement Ireland refers, amongst others, to the case *ex parte* Alliance against Birmingham Northern Relief Road: "the fact that that Agreement can be described as a commercial document does not mean that it does not contain information which relates to the environment".

40. Counsel for the United Kingdom, Transcript, Day 2 Proceedings, pp. 68-69.

41. *Id.*, p. 86.

63. With respect to the hazardous nature of the MOX production Ireland submitted –

> *MOX production is an activity which will inevitably and certainly affect the maritime area, including Ireland's waters. It will do so principally in three ways: (1) routine (intentional) discharges from MOX; (2) routine (intentional) discharges from THORP, due to the intensification of activity aimed at producing materials for the MOX plant; (3) discharges from possible accidents or terrorist attacks, either from the MOX plant itself or from transports of radioactive waste to, or MOX from, the plant.*[42]

64. The United Kingdom did not respond adequately to this contention, but rather focused its arguments on establishing the link between the Reports and future discharges and contended that –

> *the relevant question . . . is not whether the MOX production will affect the maritime area: it is whether the information requested is information on activities or measures adversely affecting or likely to affect the maritime area.*[43]

65. Hence, the issue of adverse effect under the second category of information, that has been elevated by the majority so as to emerge as decisive in the result, received only a passing reference and attention in the written submissions of the Parties and the oral pleadings.[44]

66. The majority (para. 179) finds against Ireland on the ground that –

> *In the opinion of the Tribunal, Ireland has failed to demonstrate that the Indicative List's 14 categories of redacted items in the PA and ADL Reports, insofar as they may be taken to be activities or measures with respect to the commissioning and operation of a MOX plant at Sellafield, are 'information . . . on the state of the maritime area' or, even if they were, are likely adversely to affect the maritime area.*

67. In maintaining this conclusion against the second category of information of Article 9(2) the majority summarily rejects the possibility of substantial environmental damage to the Irish Sea by –

---

42. *See* Ireland's Memorial, para. 96. *See also* United Kingdom's Counter-Memorial, para. 4.2; Counsel for Ireland, Transcript, Day 1 Proceedings, p. 46; Counsel for the United Kingdom, Transcript, Day 2 Proceedings, p. 85.

43. United Kingdom's Counter-Memorial, para. 4.3.

44. Counsel for the United Kingdom, Transcript, Day 2 Proceedings, pp. 62-63.

(1) casting an onus on Ireland to prove adverse effect that has not been discharged; and

(2) criticising Ireland for focussing on the wrong issues.

68. To my mind this is to mistake the issues for determination. Neither Party contended that the PA and ADL Reports are in themselves activities or measures with respect to the commissioning and operation of the MOX Plant. Although it may be that the Reports may in any event fall within the scope of regulatory initiatives, the primary point of dispute between the Parties was whether the Reports contained information *on* activities or measures within Article 9(2). The Parties' common position was that the Tribunal must first identify the subject activity or measure, and then consider whether a link exists between it and the Reports.

69. Albeit that the United Kingdom contended that such effect was small and broadly neutral, the Parties were agreed that the manufacture of MOX fuel may affect the maritime area.[45] Indeed, the entire justification exercise engaged by the United Kingdom was in the context that negative environmental effect arising from the commissioning of the Plant was a given, and that commissioning should not be approved unless the proven economic effect outweighed the detriment.

70. The underlying substantive disagreement between the Parties, which is not the subject of the arbitration, is whether future radioactive emissions are of such a magnitude that they may significantly harm the marine environment of the Irish Sea. It was this admitted environmental detriment that supported the Tribunal's mandate to consider the Reports and to apply the definitions of Article 9(2).

71. I conclude that it was not open to the majority to find no adverse effect as a determinative finding of fact. I contest the majority's assumption that its mandate enabled it to make this summary finding as determinative to its decision adverse to Ireland.

45. While the United Kingdom does not admit that explicitly in the proceedings, it at the same time submits that the information on potential environmental impacts of the MOX Plant has been made available to Ireland: "Ireland has known for many years what the liquid and gaseous discharges from the MOX Plant are likely to be; it has known for many years what the radiological impact of the MOX Plant is likely to be (and moreover it does not challenge the United Kingdom's estimates on radiation doses from the MOX Plant)." United Kingdom's Counter-Memorial, para. 4.7. *See also* Counsel for the United Kingdom, Transcript, Day 2 Proceedings, p. 86.

Burden of Proof

72. In my opinion the majority also is in error to assume, and to apply, the burden of proof as falling on Ireland. In this regard, I maintain that the obvious application of the precautionary principle (not considered by the majority) must shift the burden to the United Kingdom.

73. As an established customary principle of international law,[46] the precautionary principle is embraced by Article 2(2)(a) of the OSPAR Convention[47] –

> *ARTICLE 2 GENERAL OBLIGATIONS*
>
> *2. The Contracting Parties shall apply:*
>
> *a. the precautionary principle, by virtue of which preventive measures are to be taken when there are reasonable grounds for concern that substances or energy introduced, directly or indirectly, into the marine environment may bring about hazards to human health, harm living resources and marine ecosystems, damage amenities or*

---

46. While one might contest the customary nature of the precautionary principle, on this matter I am guided by the views expressed by the European Union institutions. Amongst others, in 2000 the European Commission issued a Communication on Precautionary Principle, in which it suggested that the principle "has been progressively consolidated in international environmental law, and so it has since become a full-fledged and general principle of international law". In support of its position the Commission cited international legal instruments, EU legislation as well as its own practice, and in particular: Ministerial Declaration of the Second International Conference on the Protection of the North Sea (1987); Convention on Biological Diversity (1992); Framework Convention on Climate Change (1992); Paris Convention for the Protection of the Marine Environment of the North-East Atlantic (22 September 1992) and WTO Agreement on the Application of Sanitary and Phytosanitary Measures (SPS); Resolution of the European Parliament of 10 March 1998 concerning the Green Paper on the General Principles of Food Law in the European Union of 30 April 1997, Council Resolution of 13 April 1999 and Resolution of the Joint Parliamentary Committee of the EEA (European Economic Area) of 16 March 1999 (Annex I, Refs. 8-12); Communication of 30 April 1997 on consumer health and food safety (COM(97) 183 final), Green Paper on the General Principles of Food Law in the European Union of 30 April 1997 (COM(97) 176 final). *See* European Commission Communication on the Precautionary Principle, COM(2000) 1. *See* http://europa.eu.int/comm/food/fs/ifsi/eupositions/ccgp/ccgp01_en.html Shortly after the circulation of the Communication, the Nice Council adopted a Resolution reiterating that "the precautionary principle is gradually asserting itself as a principle of international law in the fields of environmental and health protection". *See* Presidency Conclusions, Nice European Council Meeting, 7, 8 and 9 December 2000, Annex III, Council Resolution on the precautionary principle. *See*, in particular, para. 3.

47. Article 2(1)(a) of the OSPAR Convention reads: "the Contracting Parties shall, in accordance with the provisions of the Convention, take all possible steps to prevent and eliminate pollution and shall take the necessary measures to protect the maritime area against the adverse effects of human activities so as to safeguard human health and to conserve marine ecosystems and, when practicable, restore marine areas which have been adversely affected."

> *interfere with other legitimate uses of the sea, even when there is no conclusive evidence of a causal relationship between the inputs and the effects; . . .*

At the least, the precautionary principle, as expressed in Article 2(2)(a), to apply *"even when there is no conclusive evidence of a causal relationship between the import and the effects"*, directs the majority to consider the application of the principle. In my opinion it goes further and explicitly has the effect of transferring the responsibility for providing scientific evidence to the producer of hazardous substances (or, as here, to the Ministers of State as the decision-makers).

74. To adapt the European Commission, the OSPAR Convention *"by way of precaution, has clearly reversed the burden of proof by requiring that the substances be deemed hazardous until proven otherwise"*.[48] It must follow from the fact that the OSPAR Convention incorporates the precautionary principle as a General Obligation under Article 2 that on these issues arising under Article 9 it is the United Kingdom that bears responsibility for proving that future damage is insignificant and that there is no likelihood of adverse effect.

75. The finding by the majority that Ireland *"has failed to demonstrate"* adverse effect within the second category of information ignores Article 2(2)(a). For this reason, I conclude that the majority has misdirected itself on the question of onus. I regard the precautionary principle so engaged by Article 2(2)(a) as requiring that any finding adverse to Ireland be made in terms inverse to those found by the majority, namely that such a result was and is only open to be made on a finding that *in fact* there was no such potentially adverse effect. Clearly this was not a matter of fact established on the material before the Tribunal. Nor, as I understand its position, was that contended for by the United Kingdom.

76. As the majority did not consider the precautionary principle and misdirected itself on the question of onus, I conclude that its finding that Ireland *"has failed to demonstrate"* adverse effect within the second category of information must be vitiated as predicated upon the wrong approach to the burden of proof.

77. In any event, even were the burden of proof on this issue on Ireland (as the majority assumes), in focussing on the nature of the link between the Reports and hazardous activity Ireland was primarily refuting the United Kingdom's argument cited above, namely –

---

48. Communication from the EC Commission on the precautionary principle. COM(2000) 1, para. 6.4. *See also* the following observation: "the main effect of the [precautionary] principle . . . is to require states to submit proposed activities affecting the global commons to international scrutiny." Patricia W. Birnie and Alan E. Boyle, International Law and the Environment, at 98 (1995).

> *The relevant question, however, is not whether MOX production will affect the maritime area. It is whether the* information requested *is information on activities or measures adversely affecting or likely to affect the maritime area . . . .*[49]

Further,

> *[T]he material words of provision cover only information which is directly and proximately related to the state of the maritime area or to activities or measures adversely affecting or likely to affect the maritime area. On first impression, the information sought by Ireland fails to meet this test.*[50]

78. Were it relevant, the fact that it appears to the majority that Ireland had not addressed the issue of adverse effect adequately could not be used as a self-standing legal argument in favour of ruling out the possibility of significant environmental harm. Were that the case, the Tribunal should have ensured that the Parties were given an opportunity to present their positions in a comprehensive way by further submissions, as on the third category issue which is the subject matter of *Decision No. 5*. The Tribunal was not in a position to determine the matter on the basis of the submissions made up to the end of the hearings, and should not have done so.

Likely Adverse Effect

79. As a further matter of dissent, I disagree with the interpretation of the adverb *"likely"* propounded by the majority in paragraph 175. That finding suggests that its inclusion in Article 9(2) creates an additional, and therefore higher, threshold for qualifying negative impacts. In my view, it signifies the opposite. This is because the definition of Article 9(2) does not instruct the Tribunal to find that it is established that there *will be* significant detriment to the Irish Sea. It speaks only of *potentially* adverse effect.

---

49. *See* United Kingdom's Counter-Memorial, para. 4.3.

50. *See* United Kingdom's Counter-Memorial, para. 4.8. In its oral representations Ireland summarised the main arguments employed by the United Kingdom: "What does the United Kingdom say? It makes three arguments. Firstly, the information is of a purely commercial character; secondly, the information is not directly and proximately related to activities or measures adversely affecting or likely to affect the maritime area and, thirdly, Ireland's approach is based on the *Mecklenburg* case of the European Court of Justice, which is not on point, and the Aarhus Convention which is not applicable in the exercise of the progressive development which is not in force." Counsel for Ireland, Transcript, Day 1 Proceedings, p. 46. Wholly aside from the question whether this summary is correct, Ireland perceived the United Kingdom's position in the way described and focused its defence on refuting the above submissions.

80. The difference between these two approaches is self-evident. In the first case, definite and unavoidable harm would have to be shown as a fact. In the second case, it suffices to demonstrate that significant damage was probable.

81. In this context *"likely"* in its ordinary meaning means *"probable"* as something expected but not certain to happen.[51] The result of *"likely adverse effect"* is not a fact that is required to be proven empirically, but is merely to be recognised as a possibility as something that *may*, but not necessarily *will* happen. Perhaps the phrase *"reasonably to be expected"* accurately expresses the standard.

82. In other words, when qualifying an activity as potentially harmful the Tribunal must be guided by the word *"likely"* that applies a lower threshold of proof for satisfaction at the level that it is not *"adverse effect"* that must be established, but merely the likelihood of such adverse effect.

Findings of Fact

83. A further criticism against the majority's approach on these issues of findings of fact is that it fails to acknowledge the general legal context in which the dispute is being resolved. To my mind, the majority has impermissibly made its summary determination adverse to Ireland as an unsubstantiated judgement on the principal issue raised by the Parties in disregard of their submissions.

84. The issue of adverse effect lies at the heart of the various legal actions pursued by Ireland against the United Kingdom. The environmentally adverse character of the MOX Plant operation is subject to parallel sets of proceedings. In its oral submissions[52] the United Kingdom referred to the fact that the issue has been extensively argued between the same Parties in the International Tribunal for the Law of the Sea ("ITLOS") proceedings in November 2001.[53] The Tribunal in those proceedings has not pronounced on the question of future environmental detriment,[54] but has ordered the Parties to co-operate fully, and in particular to –

---

51. Oxford English Dictionary, 2nd ed.

52. Counsel for the United Kingdom, Transcript, Day 2 Proceedings, pp. 62-63.

53. *The Mox Plant Case (Ireland v. United Kingdom)*, Request for Provisional Measures, Order Dated December 3, 2001, International Tribunal for the Law of the Sea, Case No. 10. The Order of the Tribunal and the transcript of the Parties' written and oral submissions can be accessed on the ITLOS web site at http://www.itlos.org. Oral hearings took place in Hamburg on 19 and 20 November 2001.

54. This particular issue was not within its mandate.

> (a) *exchange further information with regard to possible consequences for the Irish Sea arising out of commissioning of the MOX plant;*
>
> (b) *monitor risks or the effects of the operation of the MOX plant for the Irish Sea;*
>
> (c) *devise, as appropriate, measures to prevent pollution of the marine environment which might result from the operation of the MOX plant.*[55]

85. This conclusion of the ITLOS indicates that the question of future adverse effect is still open. The mandate of the arbitral tribunal established in accordance with Annex VII of the United Nations Convention on the Law of the Sea ("UNCLOS") is defined by Ireland as –

> *to address the dispute which concerns '(a) pollution of the Sea arising from operation of the plant and (b) the risks arising from movements of material to and from the plant, and relates to the failure of the United Kingdom (1) to co-operate with Ireland, (2) to protect the marine environment and (3) to take all necessary measures to prevent, reduce and control pollution.'*[56]

The matters before the UNCLOS Tribunal remain unresolved following its June 2003 hearings.

86. In these continuing circumstances of disputation, at the least the issue joined between the United Kingdom and Ireland as to the potentially adverse environmental harm is undecided. The evidence is inconclusive. The Parties continue exchanging evidence to that effect, and the debate at the European Union level over the consequences of the commissioning of the MOX Plant continues.[57]

---

55. *Supra* note 53, dispositif, para. 1.

56. Official statement by Joe Jacob, TD, Minister with responsibility for Nuclear Safety, of 26 October 2002, citing a Notification by Ireland of the "Dispute Concerning the MOX Plant, International Movements of Radioactive Materials, and the Protection of the Marine Environment of the Irish Sea" with a Statement of Claim and Grounds upon which it is based. Published at http://www.irlgov.ie/tec/press01/october26th01.htm.

57. *See*, for instance, Commission Opinion of 26 November 2002 concerning the plan for the disposal of radioactive waste arising from the operation of the MOX Demonstration Facility at Sellafield located in the United Kingdom, in accordance with Article 37 of the EURATOM Treaty, 2002 OJ (C 292), pp. 0007-0008. *See also* Written Question E-0649/02 by European Parliament, Nuala Ahern (Verts/ALE), to the Commission on the subject of radioactive discharges, 2002 OJ (C 229), pp. 0113-0114.

87. In contrast with the fact of the majority making a determinative finding on this perceived issue of fact, before the Tribunal the Parties more focused their submissions on interpreting the inclusiveness of the definition and proving whether the link exists between the Reports and the commissioning of the MOX Plant. I regard this issue of enquiry identified by the Parties as the primary task for this Tribunal.

88. I characterise the majority's conclusion against Ireland on the adverse effect issue as an unsupported assertion made without any reference to the materials of the case. The Majority Opinion considers none of the matters militating to the contrary contained in the examination of future environmental detriment advanced in Ireland's Memorial,[58] Reply,[59] or oral submissions,[60] or the references to the issue of adverse effect made by the United Kingdom in the course of oral pleadings.[61]

89. Hence, I regard this summary finding by the majority against the likelihood of adverse effect to the Irish Sea as made without any assessment and identification of its factual basis, and in apparent disregard of the admitted environmental damage, the lower threshold of proof and the long-standing dispute over the threat to the marine environment associated with the MOX Plant.

90. As the United Kingdom has not established the fact of no adverse effect (as required by the application of the precautionary principle) it is unnecessary to make a finding. In any event, and were it a matter for determination, I am inclined to conclude that the material adduced by Ireland militates in favour of the probability of substantial environmental damage. The 1993 MOX Plant Environmental Impact Statement (which has never been revised or updated) must be read as premised on the view that the Plant may adversely affect the environment.[62] Each of the 1998 Proposed Decision on Justification[63] and the 2001 DEFRA Decision extensively analyses environmental impacts and safety

58. *See*, amongst others, Ireland's Memorial, paras. 19-23 and 96-97.

59. *See*, for instance, Ireland's Reply, paras. 12-13.

60. *See*, amongst others, Counsel for Ireland, Transcript, Day 1 Proceedings, pp. 61-62.

61. Counsel for the United Kingdom, Transcript, Day 2 Proceedings, pp. 62-63.

62. Ireland's Memorial, Annex 9.

63. *See* paras. 22-31 of the Proposed Decision. Ireland's Memorial, Annex 5.

concerns related to the Plant[64] and emphasises that there will be radioactive discharges into the marine area of the Irish Sea.[65] Such waste inherently has the potential to harm marine life and damage human health. Further, the 1998 Sintra Ministerial Decision, signed by the United Kingdom, requires that –

> *discharges, emissions and losses of radioactive substances are reduced by the year 2020 to levels where the additional concentrations in the marine environment above historic levels, resulting from such discharges, emissions and losses, are close to zero.*[66]

91. However, in my opinion such matters of likely fact may be put on one side, as I regard the issue of adverse effect as one with respect to which this Tribunal does not have competence and one which is not open to the majority to determine as a fact dispositive of these proceedings.

92. In summary, and for these reasons, my objections against the majority interpretation of the extent and meaning of adverse effect in Article 9(2) are that –

(1) it fails to address the admitted environmental harm to the marine environment of the Irish Sea, as well as the fact that Article 9(2) only speaks of the likelihood of adverse effect. These two factors create a lower threshold of proof for Ireland;

(2) in accordance with the precautionary principle, the burden of proof lies with the United Kingdom;

(3) the majority conclusion appears to be unfounded, since no factual evidence was presented in support of its finding; and

(4) the available material militates in favour of the conclusion that the probability of adverse effect might be demonstrated.

---

64. *See* Decision of the Secretary of State for Environment, Food and Rural Affairs and the Secretary of State for Health of October 3, 2001, paras. 56-70. Ireland's Memorial, Annex 5.

65. *See* paras. 56-70 of the 2001 Decision. Ireland's Memorial, Annex 5.

66. *See* the 1998 Sintra Ministerial Statement. Ireland's Memorial, Annex 8.

Relationship Between the Reports and Activities Likely Adversely to Affect the Marine Environment

93. The defining issue for the Tribunal's consideration is whether a link exists between the harmful activity and the information contained in the PA and ADL Reports.

94. In answering this question, the majority invokes and considers what it calls the *"theory of inclusive causality"*. The majority maintains that *"under an interpretative theory of inclusive causality, anything, no matter how remote, which facilitated the performance of an activity is to be deemed part of that activity"* (para. 164). This theory is then picked up as the vehicle for denunciation of Ireland's arguments in the following paragraphs of Part XI, with the majority concluding (para. 174) that because Article 9(2) establishes an additional threshold of adversity, the concept of *"inclusive causality"* fails.

95. The sources and definition of this *"theory of inclusive causality"* embraced by the majority are unknown to me. The expression was not part of either Party's submissions, and it conveys nothing beyond what I discern from the majority's references to it. I have had a nil return on my searches of textbooks on international law, Lexis-Nexis, a bibliography on causation,[67] dictionaries and the like, and I have not found any reference to the semantic linking of the words *"inclusive"* and *"causality"*. In explanation, it may be that the majority is doing no more than attaching to Ireland's arguments their own creative appellation. If so, such a nomenclature in no way assists in the consideration of the case.

96. Plainly, the PA and ADL Reports are a cause (or, more correctly, one of the causes) necessary for a harmful activity to occur, in the sense that at least the following are demonstrated –

- the submission of the Reports was causally prior to the occurrence of the harmful activity;
- the commissioning of the Reports and the actual harmful activity are two distinct factual events; and that
- in the circumstances, had the Reports not been provided, the harmful activity would not be authorised or occur.

---

67. *See* Partial Bibliography on Causation, compiled by Ellery Eels and Dan Hausmann, published at http://www.vanderbilt.edu/quantmetheval/causality.htm.

97. I support the United Kingdom's proposition that information falling within the scope of Article 9(2) is required to concern a harmful activity and its effect on the state of the maritime area.[68] However, and contrary to its submissions,[69] I conclude that Article 9(2) does not require this link to be direct and proximate,[70] or even sufficiently proximate.[71] In my opinion, Article 9(2) merely requires the existence of *any* relationship between future negative effects of the MOX Plant operation and the information contained in the PA and ADL Reports. In this regard, the justification exercise carried out by the United Kingdom is of primary importance to establishing this link.

98. The United Kingdom does not contest its obligation to justify the operation of the MOX Plant by a process that *"requires a consideration of whether the benefits of the practice outweigh the detriments"*.[72] It also agreed with Ireland that the PA and ADL Report processes were carried out *"within the context of the justification exercise under Euratom Directive 96/29 (replacing Directives 80/836 and 84/467)"*.[73]

99. Ireland contended that the obligation to justify requires an identification of the economic costs and benefits of the MOX Plant operation, and that, because of the omission of the economic data from the public domain versions of the PA and ADL Reports, Ireland is unable to assess whether potential negative environmental consequences of the MOX Plant operations are justified from the economic standpoint.[74]

---

68. Counsel for the United Kingdom, Transcript, Day 1 Proceedings, p. 92. United Kingdom's Counter-Memorial, para. 4.12.

69. *See*, for instance, the following statement: "the material words of that provision [Article 9(2)] cover only information which is directly and proximately related to the state of the maritime area or to activities or measures adversely affecting or likely to affect the maritime area." United Kingdom's Counter-Memorial, para. 4.8.

70. Ireland maintains that even though Article 9(2) does not contain such a condition, both the PA and ADL Reports meet the requirement of directness and proximity. *See* Counsel for Ireland, Transcript, Day 1 Proceedings, pp. 49-50.

71. United Kingdom's Counter-Memorial, para. 4.12.

72. *Id.*, footnote 5, at pp. 3-4.

73. *Id.*, para. 1.3.

74. Ireland's Memorial, para. 40; Counsel for Ireland, Transcript, Day 1 Proceedings, p. 11 and Day 4 Proceedings, p. 66. *See also* Letter from Renee Dempsey to Michael Wood, 7 August 2001, para. 13. *See also* the Second MacKerron Report which reads: "This is an admission that without the information sought, the economic case for the SMP cannot be assessed. This goes contrary to Article 6 of the Directive 80/836/EURATOM and Article 6 of Directive 96/269. Ireland, who has a material interest in

100. To this extent, it appears that the Parties were agreed that the assessment of benefits, including economic benefits, constitutes an integral part of the justification exercise. They diverged on the issue whether information concerning such benefits in the assessment process may fall within the scope of the definition of Article 9(2).

101. The United Kingdom contended that the information excised from the PA and ADL Reports was of a purely commercial nature[75] and that each of the PA and ADL Reports was an independent review of the business case for the commissioning of the MOX Plant.[76] It submitted that the information ceased to be relevant to the environment[77] upon the Executive concluding that the balance was broadly neutral.

102. Although Ireland accepted that the redacted information related to the commercial activity of the operation of the MOX Plant,[78] it submitted that, nonetheless, such information directly affected the environment[79] and that its commercial character was not determinative of characterisation as information within Article 9(2).[80]

103. Ireland identified that the *"purpose of the PA and ADL Reports is to examine the justification of the MOX Plant, taking into account inter alia the economic costs of its environmental consequences and of measures taken to limit those*

---

the environmental consequences of the SMP, is unable to assess without the information sought, whether there ever was an economic justification to the SMP. The statement by David Wadsworth confirms this." (Appendix B, para. B.1.1). Ireland's Reply, para. 34.

75. United Kingdom's Counter-Memorial, para. 4.10; Counsel for the United Kingdom, Transcript, Day 2 Proceedings, p. 67.

76. Counsel for the United Kingdom, Transcript, Day 2 Proceedings, p. 96.

77. *See*, for instance, the following statement: "given its conclusion on environmental and other issues, that the balance is broadly neutral, the draft decision then went on to consider the economic case concluded that there was a case for approval. This is the point in the stage of consultations at which consideration of the environmental issues was concluded and from this point onwards, essentially, the issues being considered are no longer environmental, . . . the issues considered hereafter were the commercial arguments for and against the plant or the process." Counsel for the United Kingdom, Transcript, Day 2 Proceedings, pp. 68-69.

78. Counsel for Ireland, Transcript, Day 1 Proceedings, p. 48.

79. *Supra* note 38.

80. *Supra* note 39.

environmental consequences".[81] It further submitted that the redacted commercial data is directly related to the environment because it sheds light on –

> *. . . whether all the costs have been properly integrated into the design and operation of the plant; whether best environmental practices are being budgeted for; whether best available technology is being used; whether best available technology will continue to be used in the coming years as technology evolves.*[82]

104. The United Kingdom was required to "justify" the MOX Plant under the applicable approval processes before its operation may be authorised. This relevant obligation first imposed on the United Kingdom in 1980 by Directive 80/836/EURATOM, Article 6, which provided –

> *. . . the limitation of individual and collective doses resulting from controllable exposures shall be based on the following general principles: (a) every activity resulting in an exposure to ionising radiation shall be justified by the advantages which it produces,*[83]

was replaced in 1996 by Directive 96/29/EURATOM, Article 6(1) in terms –

> *Member States shall ensure that all new classes or types of practice resulting in exposure to ionising radiation are justified in advance of being adopted by their economic, social or other benefits in relation to the health detriment they may cause.*[84]

105. The applicability of these 1996 EURATOM standards to the "justification test" is accepted by the Parties.[85]

---

81. Written outline of submissions on behalf of Ireland, on file at the offices of the PCA (Professor Sands), para. 62.

82. Counsel for Ireland, Transcript, Day 1 Proceedings, p. 26. *See also* the following statement: "the very purpose of the PA and ADL reports is to examine the justification of the MOX plant taking into account all economic costs and those economic costs include the cost of environmental consequences, include the costs of ensuring against environmental damage, include the costs of ensuring against transport accidents, include the costs of ensuring that the plant is safe and complies with all domestic and international environmental standards." Counsel for Ireland, Transcript, Day 1 Proceedings, p. 46.

83. Ireland's Memorial, para. 37; United Kingdom's Counter-Memorial, para. 1.13; Counsel for Ireland, Transcript, Day 1 Proceedings, p. 6; Counsel for the United Kingdom, Transcript, Day 2 Proceedings, p. 64.

84. Ireland's Memorial, para. 38.

85. *See*, for instance, United Kingdom's Counter-Memorial, paras. 1.3 and 1.11.

106. The intention and purpose of the United Kingdom to treat economic data as having direct relevance to the environment is discerned from the contents of the DEFRA Decision ("the Decision") on the justification of the MOX Plant, adopted on 3 October 2001,[86] and, in particular, from the circumstances that –

- the essence of the obligation of justification is described by the Decision as: *"the requirement of justification is based on the internationally accepted principle of radiological protection that no practice involving exposure to radiation should be adopted unless it produces sufficient benefits to the exposed individuals or to society in general to offset radiation and any other detriment it may cause"*;[87]

- the Decision states that *"the application of the justification test requires the consideration of environmental, safety, economic, social and other benefits and disbenefits"*;[88]

- the Decision explicitly relies on the EURATOM Regulations as grounds for conclusions regarding the justification of the MOX Plant:[89] paragraph 91 states that "*The Secretaries of State have concluded that the manufacture of MOX fuel is justified in accordance with the requirements of Article 6(1) of Directive 96/26/EURATOM*"; and

- paragraphs 56-70 of the Decision extensively analyse environmental detriments that may be caused by the manufacture of MOX fuel as well as safety and security concerns associated with the Plant's operation. These detriments are further balanced against economic and other benefits in paragraphs 71-81. This balance, of environmental concerns *vis-à-vis* future profits, is based primarily on the calculations produced by ADL in its Report.[90]

---

86. *Supra* fn. 64; United Kingdom's Counter-Memorial, paras. 2.22-2.23; Counsel for the United Kingdom, p. 73.

87. *Supra* fn. 64, para. 13.

88. *Id.*, para. 28.

89. *See supra* fn. 64, paras. 13-20.

90. *See*, in particular, paras. 76-77, 85 and 89 of the DEFRA Decision.

107. I demur to the contention of the United Kingdom that each of the PA and ADL Reports has nothing to do with the state of the maritime area embraced by the Convention. This is because each was commissioned by the United Kingdom Government in the framework of the mandated justification exercise considering whether economic benefits offset environmental harm.[91] This balancing process was acknowledged in the United Kingdom's Counter-Memorial as requiring *"a consideration of whether the benefits of the practice outweigh the [environmental] detriments"*.[92]

108. The significance of the environmental factors during the economic analysis is further confirmed by the explicit language of the ADL Report –

> [the Plant] cannot operate without passing a test of justification: *the benefits of a practice involving ionising radiation need to outweigh any environmental or other detriments.*[93]

109. The economic data collected and presented in the PA and ADL Reports was an integral and necessary part of the required process to determine whether the pollution of the marine environment might be legitimised under the nuclear regimes. It was this data that was deployed by the decision-makers, (at the executive level of Ministers of State) in the justification exercise for the commissioning of the MOX Plant.

110. At this point, the interdependence between economic data and environmental impacts becomes evident. It is inherent in the justification test that economic analyses may be determinative of whether future environmental harm is legitimate and whether the activity that is likely adversely to affect the maritime area should be authorised. Without economic data the exercise of justification becomes meaningless, as the second integral part of the entire test (namely the economic, social, and any other benefits in justification) will be missing.[94]

111. It is the economic analyses that provide the balancing factor to the scales of assessment in a justification process calibrated by the EURATOM Regulations to favour the environment. In the terms of physics, the moment tilting the balance

---

91. *Supra* fn. 64, para. 13.

92. United Kingdom's Counter-Memorial, footnote 5, at pp. 3-4.

93. ADL Report, para. 1.

94. In his testimony, Mr. MacKerron alleges that "justification has been established in prior cases as amounting to net economic advantage which should outweigh any radiological betterment". *See* Testimony of Mr. Gordon MacKerron, Transcript, Day 2 Proceedings, p. 6.

against approval can only be offset by a larger moment arising from the justification exercise. That was the inherent function of the Reports. As information so directly integral to the process of assessment, such information must be characterised as bearing a most direct relevance to the state of the marine environment of the North-East Atlantic. To my mind, it would be futile for the exercise, and also confound the purpose of the justification regime, to qualify by unexpressed limitations the broad definition of information under Article 9(2) (as has the majority) to enable access only to the purely environmental side of the balance and to exclude the information taken into account on the other side of the scale.

112. For these reasons, the fact that the Parties are agreed that the PA and ADL Reports are comprehensive reviews of the business case[95] and contain commercial information,[96] cannot in itself exclude the relevance of the Reports to an activity which is likely to adversely affect the maritime area.

Other Considerations

113. Ireland supported its non-restrictive interpretation of Article 9(2) as consistent with international and domestic law and practice[97] arising from the Aarhus Convention[98] as confirming the validity of the approach taken by Ireland in its broad definition.[99] It also relied upon cited ECJ jurisprudence,[100] EC legislative proposals and domestic sources.[101]

---

95. Counsel for the United Kingdom Transcript, Day 2 Proceedings, p. 96.

96. Counsel for Ireland, Transcript, Day 1 Proceedings, p. 48.

97. Ireland's Memorial, para. 100. Counsel for Ireland, Transcript, Day 1 Proceedings, pp. 55-57.

98. Article 2(3) of the Aarhus Convention reads as follows:

> 'Environmental information' means any information in written, visual, aural, electronic or any other material form on: . . . (b) Factors, such as substances, energy, noise and radiation, and activities or measures, including administrative measures, environmental agreements, policies, legislation, plans and programmes, affecting or likely to affect the elements of the environment within the scope of subparagraph (a) above, and cost-benefit and other economic analyses and assumptions used in environmental decision-making; . . .

99. Counsel for Ireland, Transcript, Day 1 Proceedings, p. 52.

100. *See, e.g.*, *Mecklenburg*, *supra* note 27, Ireland's Memorial, para. 102. Counsel for Ireland, Transcript, Day 1 Proceedings, p. 55.

101. Counsel for Ireland, Transcript, Day 1 Proceedings, pp. 56-57.

114. The United Kingdom responded that such sources were either irrelevant or inapplicable in the relations between the Parties,[102] or support the United Kingdom's position.[103]

115. Plainly, it is not the function of the Tribunal to consider or determine whether the Aarhus Convention and EC legislation reflect a progressive development of international environmental laws and regulations.[104] Nor to engage in the interpretation of such instruments. The relevant enquiry is whether the terms of these instruments relevantly and permissibly inform the proper interpretation of Article 9(2). In this regard, Article 2(3)(b) of the Aarhus Convention expressly includes in the definition of environmental information *"cost-benefit and other economic analyses and assumptions used in environmental decision-making"*.

116. For the reasons stated, I have concluded that the Convention possesses normative and evidentiary value and should be included in the complex of rules of international law in accordance with which the Tribunal is required to resolve the dispute at bar.

117. Further and beyond the Aarhus Convention, other of the materials put before the Tribunal by the Parties confirm regional trends exposing the intention of States and the European Union itself to include economic analyses in the definition of environmental information.[105] At the least, these trends appear to broaden the

---

102. United Kingdom's Rejoinder, para. 21. Counsel for the United Kingdom, Transcript, Day 2 Proceedings, p. 51.

103. *See*, for instance, the United Kingdom's interpretation of the conclusions reached by the ECJ in the *Mecklenburg* case. United Kingdom's Counter-Memorial, para. 4.11.

104. In the UK's view, the definition of the Aarhus Convention reflects "an exercise of progressive development of the law relating to 'environmental information'". United Kingdom's Counter-Memorial, para. 4.13.

105. *See*, in particular, Counsel for Ireland, Transcript, Day 1 Proceedings, pp. 68-71; United Kingdom's Counter-Memorial, para. 4.13. *See also* "Report from the Commission to the Council and the European Parliament on the experience gained in the application of Council Directive 90/313/EEC on freedom of access to information on the environment" (COM(2000) 400 final). *See* Counsel for the United Kingdom, Transcript, Day 2 Proceedings, p. 87; Outline of written submissions on behalf of the United Kingdom (Mr. Wordsworth); the "Proposal for a Directive of the European Parliament and of the Council on public access to environmental information" (COM(2000) 402 final. Counsel for the United Kingdom, Transcript, Day 2 Proceedings, p. 88); "Amended Proposal for a Directive of the European Parliament and of the Council on public access to environmental information" (COM(2001) 303 final); Common Position "with a view to adopting Directive 2002/. . ./EC . . . on public access to environmental information and repealing Council Directive 90/313/EEC" (Counsel for the United Kingdom, Transcript, Day 2 Proceedings, pp. 88-90); DEFRA "Proposals for a revised public access to environmental information consultation paper" (Ireland, Authorities Bundle 1, Tab 5, paras. 13-15. *See also* Counsel for Ireland, Transcript, Day 1 Proceedings, pp. 52-53. *But see* Oral Pleadings, Counsel for the United

content of the definition of environmental information so as explicitly to include cost-benefit and other economic analyses.

118. Be that as it may, more significantly for the issues of definition arising under the OSPAR Convention, the EU and United Kingdom proposals noted in the previous paragraph refer to the *clarification* of the existing formulation of Directive 90/313 rather than the adoption of a new definition.[106] This "clarification" approach also is evident from the unambiguous language of the 2002 EC Common Position[107] expressed by the EU Commission as based on the experience gained by Member States in the operation of Directive 90/313.[108]

119. In its ordinary sense, such a "clarification" does not so much extend meaning but merely confirms the content of existing meaning, including in the context of definitions within Article 9 of the OSPAR Convention.

120. Directive 2003/4 on public access to environmental information entered into force on 14 February 2003, to replace Directive 90/313,[109] and confirms that its main purpose is to bring EU law in compliance with the Aarhus Convention (para. 5) and environmental information is defined as –

> *The definition of environmental information should be* clarified *so as to encompass information in any form on the state of the environment, on factors, measures or activities affecting or likely to affect the environment or designed to protect it,* on cost-benefit and economic analyses used within the framework of such measures *or activities and also information on the state of human health and safety, including the contamination of the*

Kingdom, Transcript, Day 2 Proceedings, p. 88).

106. *See*, for instance, para. 14 of the DEFRA "Proposals for a revised public access to environmental information consultation paper": "*the definition of environmental information is clarified to refer specifically to the atmosphere, landscape, biological diversity etc. It is also defined to include cost benefit economic analyses and other assumptions used in the decision making process*". Para. 15 further states that "*these are minor changes. They are not expected to broaden the practical application of the regime*".

107. Para. 10 of the Common Position reads: "The definition of environmental information should be clarified so as to encompass . . ." Common Position "with a view to adopting Directive 2002/. . ./EC . . . on public access to environmental information and repealing Council Directive 90/313/EEC" (Counsel for the United Kingdom, Transcript, Day 2, p. 90).

108. "Report from the Commission to the Council and the European Parliament on the experience gained in the application of Council Directive 90/313/EEC on freedom of access to information on the environment", COM(2000) 400 final. *See* Counsel for the United Kingdom, Transcript, Day 2 Proceedings, p. 88; Outline of written submissions on behalf of the United Kingdom (Mr. Wordsworth).

109. *Supra* fn. 23.

> *food chain, conditions of human life, cultural sites and built structures in as much as they are, or may be, affected by any of those matters.*[110] *(emphasis added)*

121. Again, it is the language of clarification, rather than of extension or substitution, that is invoked. As Article 9(2) is nearly identical to the language of the old Directive, at the least the drafters of the Directive 2003/4 must be taken to have understood the definition of Article 9(2) in the same way. Hence, as much as for directive 90/313 that it replaced, Directive 2003/4 thereby appears also relevant to understanding the meaning of "information" under the Article 9(2) definition by making explicit that which already was implicit under the OSPAR Convention.[111]

122. It follows that international environmental regulations as binding as Directive 2003/4 now define that facts similar in nature to the PA and ADL Reports fall within the definition of disclosable information. Since Directive 2003/4 presently is binding on the United Kingdom, it is not amenable to be characterised as soft law or progressive development of law that is irrelevant to the interpretation of Article 9(2). And, contrary to the United Kingdom's view,[112] there is now a clear consensus and practice with the European Union as to the meaning and application of the Aarhus Convention's terms.

123. I regret that I have not had the advantage of the submissions of the Parties as to the relevance of Directive 2003/4 that came into force during the pendency of this Tribunal's award. My conditional conclusion is that Directive 2003/4 is confirmatory (but certainly not determinative) of the interpretation of Article 9(2).

Conclusion on Second Category of Information

124. For the reasons stated, I am of the opinion that the production of MOX fuel falls within the scope of the second category of Article 9(2) to the extent that it constitutes an activity which has the potential of adversely affecting the maritime area of the Irish Sea, and that, properly characterised, the entire balancing process, including consideration of the economic case in justification, is an enquiry concerning information within the terms of Article 9(2).

---

110. *Id.*, para. 10.

111. Ireland's Memorial, para. 100. Counsel for Ireland, Transcript, Day 1 Proceedings, p. 52. Counsel for the United Kingdom, Transcript, Day 2 Proceedings, p. 89.

112. United Kingdom's Counter-Memorial, para. 4.13.

Third Category of Information

125. The Tribunal's *Decision No. 5*, to which the Parties responded, raised with the Parties the question whether the information contained in the PA and ADL Reports may fall within the third category of information within Article 9(2) as related to an activity or measure introduced in accordance with the OSPAR Convention.

126. Plainly, the three categories of information are not disjunctive, and material may fall within more than one category. For this reason it appears to me appropriate to consider whether the relevant material also constitutes information under the third category. As I am of the opinion that in any event it falls under the second category this enquiry is not essential for my dissent. However, as the majority is of the contrary view on the second category, it appears to me that the majority determination is in error and incomplete by then failing to consider the third category.

127. For information to fall under the third category a relationship has to be established between an activity or measure and a specific provision of the Convention in accordance with which such a measure has been undertaken and activity carried out.

128. In this regard, the third category does not require a direct relationship between the state of the maritime area and information on such activities or measures. I reject the United Kingdom's contention to the contrary as in that case the second and third categories would be otiose as subsumed in the first category. In effect, Article 9(2) then would be read as if it provided for *"any available information . . . on the state of the maritime area,* and in particular *information on activities or measures . . . "*. Plainly such a construction is not open to be made.

129. A different issue is whether a measure is directly related to the state of the marine environment. The examples used by the United Kingdom's response to the Tribunal's *Decision No. 5* demonstrate the link between a measure and the state of the marine area. In my view, this is not an absolute requirement for all cases. However, even if one accepts the United Kingdom's proposition that a direct relationship is required in all instances, as clearly such relationship is evident here.

130. In the context of the OSPAR Convention, I support the majority's interpretation (para. 171) of the term *"measure"* as a regulatory initiative *"by any part of the governmental apparatus of the Contracting Parties with respect to matters covered by the OSPAR Convention"*. I derive support for this meaning from the

ECJ Judgement in the *Mecklenburg* case,[113] where the ECJ has helpfully defined the term *"measure"* as an *"act linked to an individual case directed towards a specific aim and having determinative effects"*.[114] The ECJ further found that this term serves merely *"to make it clear that the acts governed by the directive [whose language, as recognised by both parties, is nearly identical to Article 9(2)] included all forms of administrative activity"*.[115]

131. The authorisation of the MOX Plant undoubtedly qualifies as a measure undertaken in accordance with the OSPAR Convention. It is a form of administrative activity exercised by the United Kingdom government which is directed towards a specific aim and designed to further the disparate objective of the OSPAR Convention. The relevant provision in accordance with which this complex of measures was introduced, is Article 2(1) of Annex I –

> *Point source discharges to the maritime area, and releases into water or air which reach and may affect the maritime area, shall be strictly subject to authorisation or regulation by the competent authorities of the Contracting Parties. Such authorisation or regulation shall, in particular, implement relevant decisions of the Commission which bind the relevant Contracting Party.*

132. The justification of the MOX Plant that the United Kingdom undertook by way of arranging public consultations and commissioning and considering the Reports clearly falls within such an authorisation. The process, including obtaining the PA and ADL Reports, was a necessary measure introduced in accordance with the OSPAR Convention designed to protect the marine environment. In its written response to the Decision Ireland has reiterated that the main reason for its enquiry was to assure itself that the authorisation of the MOX Plant was carried out in a manner consistent with the OSPAR Convention.[116] It sought information on the measures introduced in accordance with the OSPAR Convention that could be found in the Reports.[117]

---

113. *See Mecklenburg*, *supra* note 27.

114. *See* para. 18 of the Judgement.

115. *Id.*, para. 20.

116. *See* Letter from the Agent for Ireland to the Secretary of the Tribunal (February 21, 2003), footnote 4 on p. 2 providing a summary of Ireland's claims to that effect.

117. *See*, for instance, the following statement: "what is at issue here is a measure and the measure is the process of justification". Counsel for Ireland, Transcript, Day 4 Proceedings, p. 56.

133. Indeed, the United Kingdom's response to *Decision No. 5* acknowledged that Article 2(1) of Annex I constitutes a measure introduced in accordance with the Convention. It would seem to follow that the Reports would fall within the third category of information of Article 9(2) when the necessary link between the Reports and this provision is established. I disagree with the United Kingdom's contention that the Reports are unrelated to the state of the marine environment for the reason that the information contained in the PA and ADL Reports was the key factor in the authorisation by the United Kingdom of radioactive discharges into the Irish Sea. The link could not be stronger.

134. As the majority states (para. 179), the PA and ADL Reports do not necessarily have to be measures or activities themselves. Neither the United Kingdom nor, more relevantly, Ireland has claimed that they do. Rather, for the purposes of application of the third category of information it suffices to establish that the Reports contain information related to distinct measures or activities introduced in accordance with the OSPAR Convention. This must be the position here. Plainly, the PA and ADL Reports have had determinative effects on the authorisation of discharges into the maritime area by the United Kingdom government, as the findings of the Reports were used by DEFRA in preparing the Decision for the Manufacture of MOX fuel.[118]

135. For these reasons, I conclude that the information contained in the PA and ADL Reports also qualifies as measures introduced in accordance with the Convention under the third category of information of Article 9(2).

Summary

In summary, I identify the principal vitiating errors of the majority in finding that none of the redacted items was information within the definition of Article 9(2) arose from its approach of –

(1) interpreting the OSPAR Convention as if it were an isolated legal regime without regard to its context within a continuum of emerged and emerging legal instruments concerning the environment, including those in a relevant sense binding on the Parties;

(2) refusing to examine the PA and ADL Reports as a whole in light of the definition of Article 9(2) and suggesting instead to test each of the 14 categories of redacted items against the definition;

118. *Supra* fn. 64.

(3) confining, without any textual support, the second and third categories of information of Article 9(2) to being *"on the state of the maritime area"*;

(4) finding, incorrectly, that future radioactive discharges into the Irish Sea do not constitute an activity which is likely adversely to affect the state of the maritime area within the second category of information of Article 9(2);

(5) assuming, incorrectly, for the second category of information the onus of proof was on Ireland to establish that the MOX fuel production is an activity which is likely adversely to affect the maritime area;

(6) failing to characterise, for the second category of information, the justification exercise, of which the PA and ADL Reports were an integral part, as concerning the state of the maritime area; and

(7) failing adequately to examine the relevance of the third category of information of Article 9(2).

CONCLUSION

For these reasons I dissent from the dispositive conclusion of the majority accepting the United Kingdom's submissions that the whole of the redacted materials in the PA and ADL Reports is not information within Article 9(2) of the OSPAR Convention and that a final award presently should be made dismissing the claims of Ireland.

I am of the opinion that the whole of the PA and ADL Reports, including the redacted items, is information within Article 9(1) and (2) and that on such finding being made the dispute called for further hearing and consideration of the contention by the United Kingdom that Article 9(3)(d) justified the redactions made to these Reports.

Gavan Griffith
2 July 2003